Lecture Notes in Mathematics

A collection of informal reports and seminars
Edited by A. Dold, Heidelberg and B. Eckmann, Zürich

128

M. Takesaki
University of California, Los Angeles, CA/USA
and Tôhoku University, Sendai, Japan

Tomita's Theory of Modular Hilbert Algebras and its Applications

Springer-Verlag
Berlin · Heidelberg · New York 1970

Introduction

In 1967, Tomita clarified the algebraic relation between a von Neumann algebra M and its commutant M' in two unpublished papers [21] and [22], and then proved the commutation theorem for tensor products of von Neumann algebras (i.e. $(M_1 \otimes M_2)' = M_1' \otimes M_2'$). In order to study the relation between M and M' in a standard representation, (for example, a cyclic representation of M induced by a faithful normal state), he introduced two basic notions, called a generalized and modular Hilbert algebras, respectively, both being related to but different from Dixmier's quasi-Hilbert algebra [4]. See §2 for definitions. It is not very difficult to show that every von Neumann algebra is isomorphic to the left von Neumann algebra of a generalized Hilbert algebra. However, by means of generalized Hilbert algebras we see how the involution of a von Neumann algebra is twisted in a Hilbert space structure. To explain his basic idea more clearly, suppose φ_0 is a faithful normal state of a von Neumann algebra M. Then the involution: $x \in M \mapsto x^* \in M$ is not an isometry in the Hilbert space structure induced by φ_0, however it is a pre-closed operator, so that in the Hilbert space $\mathfrak{H}$ constructed via φ_0 we can consider the adjoint operator F of the pre-closed operator: $x\xi_0 \to x^*\xi_0$ where ξ_0 is the cyclic vector corresponding to φ_0. Then F is nothing but the minimal closed extension of the involution: $x'\xi_0 \to x'^*\xi_0$, $x' \in M'$, of the commutant M' of M, and the adjoint operator S of F is the minimal closed extension of the involution: $x\xi_0 \to x^*\xi_0$, $x \in M$. Therefore, we can get the polar decomposition $S = J\Delta^{\frac{1}{2}}$, $\Delta = FS$, of the involution S. The operator Δ is called the modular operator and plays a key role throughout the theory. As an algebra of analytic vectors with respect to the one parameter unitary group Δ^{it}, he could define a modular Hilbert algebra contained in the generalized Hilbert algebra $M\xi_0$ which has M and M' as its left and right von Neumann algebras respectively. Furthermore, he proved that $JMJ = M'$, where J is a unitary involution of the underlying space.

In the present lecture notes we develop Tomita's theory of modular Hilbert algebras described above in §§2-12, and then apply the theory to show the so-called KMS-boundary condition, a criterion for semi-finiteness and the Radon-Nikodym theorem. In the first of these applications it is shown that to every faithful normal positive linear functional φ_0 of a von Neumann algebra M there corresponds a unique one parameter automorphism group σ_t of M such that φ_0 satisfies the Kubo-Martin-Schwinger boundary condition with respect to σ_t. The semi-finiteness of M is characterized by the fact that the one parameter automorphism group σ_t of M is inner. Through the non-commutative Radon-Nikodym theorem, it is shown that the states satisfying the KMS-boundary condition with respect to a fixed one parameter automorphism group, are unique up to the center of the von Neumann algebra.

The author expresses his deep thanks to Professor Tomita for his agreement to publish these lecture notes before the appearance of his original paper, and also to Professor Kadison for his kind hospitality at the University of Pennsylvania. He is also indebted to his colleagues of the University of Pennsylvania, Professor Fell, Professor Effros, Professor Størmer, Professor Powers, Professor Vowden, Professor Nielsen and Professor Vesterstrøm who participated in the informal seminar discussions concerning the topics.

The author hopes that the present lecture notes will assist readers to understand Tomita's theory and to develop its fruitful applications in the future.

Content

1. Preliminaries

Throughout the whole discussion, we shall treat closed unbounded operators on a Hilbert space; we therefore first provide some notations and remarks. For detail, see Chapter XII of [6].

For each densely defined closed operator T on a Hilbert space $\mathfrak{H}$ with domain $\mathfrak{D}(T)$, we consider an inner product in $\mathfrak{D}(T)$ defined by:

$$\langle \xi | \eta \rangle_T = (\xi | \eta) + (T\xi | T\eta) ;$$

$$\|\xi\|_T = (\|\xi\|^2 + \|T\xi\|^2)^{\frac{1}{2}}, \quad \xi, \eta \in \mathfrak{D}(T) .$$

Because of closedness, $\mathfrak{D}(T)$ becomes a Hilbert space with the above inner product. Since T is the closure, as a closed operator, of the restriction $T|_{\mathfrak{M}}$ of T on a linear subspace $\mathfrak{M}$ of $\mathfrak{D}(T)$ if and only if $\mathfrak{M}$ is dense in the Hilbert space $\mathfrak{D}(T)$, we shall agree that $\mathfrak{D}(T)$ denotes the Hilbert space defined above. Let $T = UH$, $H = (T^*T)^{\frac{1}{2}}$, be the polar decomposition of T. Then $\mathfrak{D}(T)$ is identical with $\mathfrak{D}(H)$ not only as a set but also as a Hilbert space.

LEMMA 1.1. A subset $\mathfrak{M}$ of $\mathfrak{D}(T)$ is dense in $\mathfrak{D}(T)$ if and only if $(1 + H)\mathfrak{M}$ is dense in $\mathfrak{H}$.

<u>Proof</u>. Put $K = (1 + H^2)^{\frac{1}{2}}(1 + H)^{-1}$. Then K is a bounded invertible self-adjoint operator on $\mathfrak{H}$. For each $\xi \in \mathfrak{D}(T)$, put $\eta = (1 + H)\xi$. Noticing that $\|T\xi\| = \|H\xi\|$, we have

$$\|\xi\|_T^2 = \|\xi\|^2 + \|H\xi\|^2 = \|(1 + H)^{-1}\eta\|^2 + \|H(1 + H)^{-1}\eta\|^2$$

$$= ((1 + H^2)(1 + H)^{-2}\eta|\eta)$$

$$= \|K\eta\|^2 + \|K(1 + H)\xi\|^2 .$$

Therefore, the map: $\xi \in \mathfrak{D}(T) \rightsquigarrow K(1 + H)\xi \in \mathfrak{H}$ is an isometry of $\mathfrak{D}(T)$ onto $\mathfrak{H}$, so that the map: $\xi \in \mathfrak{D}(T) \rightsquigarrow (1 + H)\xi \in \mathfrak{H}$ is a homeomorphism of $\mathfrak{D}(T)$ onto $\mathfrak{H}$. Therefore, a subset $\mathfrak{M}$ of $\mathfrak{D}(T)$ is dense in $\mathfrak{D}(T)$ if and only if $(1 + H)\mathfrak{M}$ is dense in $\mathfrak{H}$, which completes the proof.

2. Modular Hilbert Algebras

DEFINITION 2.1. Let $\mathfrak{A}$ be an involutive algebra over the complex number field C with involution $\xi \in \mathfrak{A}\xi \mapsto \xi^\# \in \mathfrak{A}$. $\mathfrak{A}$ is called a <u>modular Hilbert algebra</u> if $\mathfrak{A}$ admits an inner product $(\xi|\eta)$ and a complex one-parameter group $\Delta(\alpha)$ of automorphisms of $\mathfrak{A}$, called the modular automorphisms, satisfying the following conditions:

(I) $(\xi\eta|\zeta) = (\eta|\xi^\#\zeta)$;

(II) For each $\xi \in \mathfrak{A}$, the map: $\eta \in \mathfrak{A} \mapsto \xi\eta \in \mathfrak{A}$ is continuous;

(III) The subalgebra $\mathfrak{A}^2$ of $\mathfrak{A}$, spanned by the elements $\xi\eta$ with $\xi, \eta \in \mathfrak{A}$, is dense in $\mathfrak{A}$;

(IV) $(\Delta(\alpha)\xi)^{\#} = \Delta(-\bar{\alpha})\xi^{\#}$, $\xi \in \mathfrak{A}$, $\alpha \in C$;

(V) $(\Delta(\alpha)\xi|\eta) = (\xi|\Delta(\bar{\alpha})\eta)$;

(VI) $(\Delta(1)\xi^{\#}|\eta^{\#}) = (\eta|\xi)$;

(VII) $(\Delta(\alpha)\xi|\eta)$, $\xi, \eta \in \mathfrak{A}$, is an analytic function of α on C;

(VIII) For every real number t, the set $(1 + \Delta(t))\mathfrak{A}$ is dense in $\mathfrak{A}$.

If an involutive algebra $\mathfrak{A}$ over C admits an inner product satisfying conditions (I) - (III) and the following condition (IX), then $\mathfrak{A}$ is called a generalized Hilbert algebra:

(IX) The involution: $\xi \in \mathfrak{A} \mapsto \xi^{\#} \in \mathfrak{A}$ is preclosed as a real linear operator on the real pre-Hilbert space $\mathfrak{A}$.[1)]

Suppose $\mathfrak{A}$ is a modular or generalized Hilbert algebra. Let $\mathfrak{H}$ be the Hilbert space obtained by completion of $\mathfrak{A}$. To each $\xi \in \mathfrak{A}$, there corresponds a unique bounded operator $\pi(\xi)$ on $\mathfrak{H}$ defined by $\pi(\xi)\eta = \xi\eta$, $\eta \in \mathfrak{A}$. By condition (I) and (III), $\pi(\mathfrak{A})$, the set of $\pi(\xi)$, is a non-degenerate self-adjoint algebra of operators on $\mathfrak{H}$. Let $\mathfrak{L}(\mathfrak{A})$ denote the von Neumann algebra generated by $\pi(\mathfrak{A})$. It is called the left von Neumann algebra of $\mathfrak{A}$.

LEMMA 2.1. If $\mathfrak{A}$ is a modular Hilbert algebra, then there exists a unique positive self-adjoint (not necessarily bounded)

---

1) This definition of generalized Hilbert algebras is a little modified from Tomita's original one in [22]. But Theorem 3.1 shows that our definition is equivalent to Tomita's original one.

operator Δ on $\mathfrak{H}$ such that Δ^{α} is the closure of $\Delta(\alpha)$.

Proof. By condition (V), $\{\Delta(it);\ -\infty < t < +\infty\}$ is a one-parameter group of isometries on $\mathfrak{A}$. Let $U(t)$ be a closure of $\Delta(it)$. Then $U(t)$ is a one-parameter unitary group of $\mathfrak{H}$. Because $U(t)$ is unitary and $t \to (U(t)\xi|\eta)$ is a continuous function of t for each ξ, η in $\mathfrak{A}$, $U(t)$ is a strongly continuous one-parameter unitary group. By Stone's Theorem, there is a self-adjoint operator H such that $U(t) = \exp itH$. For each complex number z, put $U(z) = \exp izH$. We claim that $U(z)$ is the closure of $\Delta(iz)$. For each $\xi, \eta \in \mathfrak{A}$, the function: $z \in C \mapsto (\Delta(z)\xi|\eta)$ is analytic. For each $\eta \in \mathfrak{H}$, take a sequence $\{\eta_n\}$ in $\mathfrak{A}$ with $\eta = \lim \eta_n$. Since $\|\Delta(z)\xi\|^2 = (\Delta(2\ \mathrm{Re}\ z)\xi|\xi)$ is a continuous function of z for every $\xi \in \mathfrak{A}$, the sequence of analytic functions: $z \mapsto (\Delta(z)\xi|\eta_n)$ converges to $(\Delta(z)\xi|\eta)$ uniformly on every compact subset of C. Hence $z \to \Delta(z)\xi$ is a weakly analytic $\mathfrak{H}$-valued function, so that it is strongly analytic. Therefore, we have

$$\Delta(z)\xi = \sum_{n=0}^{\infty} \frac{z^n}{n!} \frac{d^n}{dz^n} \Delta(z)\xi\Big|_{z=0} ,$$

where the summation is taken in the strong topology in $\mathfrak{H}$. But,

$$\frac{d^n}{dz^n} \Delta(iz)\xi\Big|_{z=0} = \frac{d^n}{dt^n} \Delta(it)\xi\Big|_{t=0} = (iH)^n\xi ,$$

so that ξ belongs to $\mathfrak{D}(\exp(izH))$ and

$$\Delta(iz)\xi = \exp(izH)\xi .$$

Hence $U(z)$ is an extension of $\Delta(iz)$. Because $(1 + \Delta(t))\mathfrak{A}$ is dense in $\mathfrak{H}$ for each real number t and $U(-it)$ is positive and self-adjoint, $U(-it)$ is the closure of $\Delta(t)$. Therefore, putting $\Delta = U(-i) = \exp H$, we get the desired positive self-adjoint operator.

We shall call Δ the <u>modular</u> operator of a modular Hilbert algebra $\mathfrak{A}$. To each real number t, there corresponds an involution: $\xi \in \mathfrak{A} \mapsto \Delta^t \xi^\# \in \mathfrak{A}$. Among them, we choose the following special two involutions $\xi \mapsto \xi^*$ and $\xi \mapsto \xi^b$:

$$\xi^* = \Delta^{\frac{1}{2}}\xi^\# \quad \text{and} \quad \xi^b = \Delta\xi^\#, \; \xi \in \mathfrak{A} . \tag{2.1}$$

Then it follows immediately that

$$(\xi|\eta) = (\eta^*|\xi^*) \; ; \tag{2.2}$$

$$(\xi|\eta) = (\eta^b|\xi^\#) \quad \text{and} \quad (\xi\eta|\zeta) = (\xi|\zeta\eta^b) . \tag{2.3}$$

The involution $\xi \mapsto \xi^*$ is extended to a reflexive conjugate linear isometry J on $\mathfrak{H}$. Put $x^T = Jx^*J$ for each bounded operator x on $\mathfrak{H}$. Then $x \in \mathfrak{B}(\mathfrak{H}) \mapsto x^T \in \mathfrak{B}(\mathfrak{H})$ is an anti-automorphism of $\mathfrak{B}(\mathfrak{H})$. The involution $\xi \mapsto \xi^b$ is called the <u>adjoint</u> involution of $\xi \mapsto \xi^\#$ and $\xi \mapsto \xi^*$ is called the <u>unitary</u> involution.

3. Generalized Hilbert Algebras

Throughout this section we suppose $\mathfrak{A}$ is a generalized Hilbert algebra. Remark that a modular Hilbert algebra is a generalized Hilbert algebra, because the adjoint involution $\xi \in \mathfrak{A} \mapsto \xi^b \in \mathfrak{A}$ is a densely defined adjoint operator of the conjugate linear operator

$\xi \mapsto \xi^{\#}$ as a real linear operator on the real Hilbert space $\mathfrak{H}$. By assumption (IX), there exists a real linear densely defined operator: $\eta \mapsto \eta^{b}$ which is the adjoint operator of $\xi \mapsto \xi^{\#}$. Let $\mathfrak{D}^{b}$ denote the definition domain of $\eta \mapsto \eta^{b}$. Then we have

$$\mathrm{Re}(\xi^{\#}|\eta) = \mathrm{Re}(\xi|\eta^{b}), \ \xi \in \mathfrak{A}, \ \eta \in \mathfrak{D}^{b}.$$

By the equation:

$$\mathrm{Re}(\xi^{\#}|i\eta) = -\mathrm{Re}(i\xi^{\#}|\eta) = \mathrm{Re}((i\xi)^{\#}|\eta)$$

$$= \mathrm{Re}(i\xi|\eta^{b}) = \mathrm{Re}(\xi|-i\eta^{b}),$$

we have

$$(i\eta)^{b} = -i\eta^{b}, \ \eta \in \mathfrak{D}^{b}.$$

Therefore, the operator; $\eta \mapsto \eta^{b}$ is conjugate linear. If $(\xi^{\#}|\eta)$ is real, then we have

$$0 = \mathrm{Re}\, i(\xi^{\#}|\eta) = \mathrm{Re}(\xi^{\#}|-i\eta)$$

$$= \mathrm{Re}(\xi|(-i\eta)^{b}) = \mathrm{Re}(\xi|i\eta^{b})$$

$$= -\mathrm{Re}\, i(\xi|\eta^{b}),$$

so that $(\xi|\eta^{b})$ is again real. For each $\xi \in \mathfrak{A}, \ \eta \in \mathfrak{D}^{b}$, take a real number θ such that $e^{i\theta}(\xi^{\#}|\eta)$ is real. Then we have

$$e^{i\theta}(\xi^{\#}|\eta) = ((e^{-i\theta}\xi)^{\#}|\eta) = (e^{-i\theta}\xi|\eta^{b}) = e^{-i\theta}(\xi|\eta^{b}),$$

so that

$$(\xi|\eta^{\flat}) = e^{2i\theta}(\xi^{\#}|\eta) .$$

Hence we get

$$(3.1) \qquad (\xi^{\#}|\eta) = (\eta^{\flat}|\xi), \ \xi \in \mathfrak{A}, \ \eta \in \mathfrak{D}^{\flat} .$$

By the equality:

$$(\xi^{\#}|\eta^{\flat}) = \overline{(\eta^{\flat}|\xi^{\#})} = \overline{(\xi^{\#\#}|\eta)} = (\eta|\xi) ,$$

$\mathfrak{D}^{\flat}$ is invariant under the map: $\eta \mapsto \eta^{\flat}$ and $\eta^{\flat\flat} = \eta, \ \eta \in \mathfrak{D}^{\flat}$. The map: $\eta \mapsto \eta^{\flat}$ will be called the adjoint involution of $\xi \mapsto \xi^{\#}$ as in the case of modular Hilbert algebras.

Because $\mathfrak{D}^{\flat}$ is the definition domain of the closed operator $\eta \mapsto \eta^{\flat}$, we consider the inner product $\langle \ | \ \rangle_{\flat}$ in $\mathfrak{D}^{\flat}$ defined by:

$$(3.2) \qquad \langle \eta_1|\eta_2\rangle_{\flat} = (\eta_1|\eta_2) + (\eta_2^{\flat}|\eta_1^{\flat}), \ \eta_1, \eta_2 \in \mathfrak{D}^{\flat} ,$$

which makes $\mathfrak{D}^{\flat}$ a Hilbert space.

Take and fix an η in $\mathfrak{D}^{\flat}$. Define operators a and b by:

$$a\xi = \pi(\xi)\eta \quad \text{and} \quad b\xi = \pi(\xi)\eta^{\flat}, \ \xi \in \mathfrak{A} .$$

Then we have, for each $\xi, \zeta \in \mathfrak{A}$,

$$(a\xi|\zeta) = (\pi(\xi)\eta|\zeta) = (\eta|\pi(\xi)^{*}\zeta)$$

$$= (\eta|\xi^{\#}\zeta) = (\zeta^{\#}\xi|\eta^{\flat})$$

$$= (\xi|\pi(\zeta)\eta^{\flat}) = (\xi|b\zeta) ;$$

hence $a^{*} \supset b$ and $b^{*} \supset a$. Therefore a and b are both preclosed. Let $\pi'(\eta)$ and $\pi'(\eta^{\flat})$ denote their closure respectively.

LEMMA 3.1. Both $\pi'(\eta)$ and $\pi'(\eta^{\flat})$ commute with every operator of $\mathfrak{L}(\mathfrak{A})$, that is, $\pi'(\eta)$ and $\pi'(\eta^{\flat})$ are affiliated with $\mathfrak{L}(\mathfrak{A})'$.

<u>Proof</u>. Because of symmetry, it is sufficient to prove the assertion only for $\pi'(\eta)$, which is equivalent to proving that $\pi'(\eta)^*$ is affiliated with $\mathfrak{L}(\mathfrak{A})'$. Take an arbitrary element ζ in the definition domain $\mathfrak{D}(\pi'(\eta)^*)$ of $\pi'(\eta)^*$. For each $\xi_1, \xi_2 \in \mathfrak{A}$, we have

$$(\pi'(\eta)\xi_1 | \pi(\xi_2)\zeta) = (\pi(\xi_2)^*\pi(\xi_1)\eta | \zeta)$$

$$= (\pi(\xi_2^{\#}\xi_1)\eta | \zeta) = (\pi'(\eta)\xi_2^{\#}\xi_1 | \zeta)$$

$$= (\xi_2^{\#}\xi_1 | \pi'(\eta)^*\zeta) = (\xi_1 | \pi(\xi_2)\pi'(\eta)^*\zeta) ,$$

so that $\pi(\xi_2)\zeta$ belongs to $\mathfrak{D}(\pi'(\eta)^*)$, and we have $\pi'(\eta)^*\pi(\xi_2)\zeta = \pi(\xi_2)\pi'(\eta)^*\zeta$. Now, take an arbitrary $x \in \mathfrak{L}(\mathfrak{A})$. Since $\pi(\mathfrak{A})$ is a strongly dense $*$-subalgebra of $\mathfrak{L}(\mathfrak{A})$, we can find a sequence $\{\xi_n\}$ in $\mathfrak{A}$ with

$$\lim_{n\to\infty} \pi(\xi_n)\zeta = x\zeta ;$$

$$\lim_{n\to\infty} \pi(\xi_n)\pi'(\eta)^*\zeta = x\pi'(\eta)^*\zeta .$$

Then we have

$$\lim \pi'(\eta)^*\pi(\xi_n)\zeta = \lim \pi(\xi_n)\pi'(\eta)^*\zeta$$

$$= x\pi'(\eta)^*\zeta ;$$

hence $x\zeta$ belongs to $\mathfrak{D}(\pi'(\eta)^*)$, and

$$\pi'(\eta)^* x\zeta = x\pi'(\eta)^*\zeta$$

by the closedness of $\pi'(\eta)^*$. Thus, $\pi'(\eta)^*$ commutes with x. This completes the proof.

DEFINITION 3.1. If $\pi'(\eta)$, $\eta \in \mathfrak{D}^\flat$, is bounded, then η is called π'-<u>bounded</u>. Let $\mathfrak{A}'$ denotes the set of all π'-bounded elements.

Remark that if η is π'-bounded, then $\pi'(\eta)^* = \pi'(\eta^\flat)$.

For each $\xi \in \mathfrak{H}$ and $\eta \in \mathfrak{A}'$, define a product of ξ and η by:

$$\xi\eta = \pi'(\eta)\xi \quad (3.3)$$

Since $\pi'(\eta)$, $\eta \in \mathfrak{A}'$, commutes with $\pi(\xi)$, $\xi \in \mathfrak{A}$, the product defined by (3.3) is compatible with the original product in $\mathfrak{A}$.

LEMMA 3.2. $\mathfrak{A}'$ is an involutive algebra with involution $\eta \mapsto \eta^\flat$ and the just defined product, and π' is an anti-representation of $\mathfrak{A}'$ on $\mathfrak{H}$.

<u>Proof</u>. Take any two elements η_1 and η_2 in $\mathfrak{A}'$. Then we have, for each $\xi \in \mathfrak{A}$,

$$\begin{aligned}(\xi|\eta_1\eta_2) &= (\xi|\pi'(\eta_2)\eta_1)\\ &= (\pi'(\eta_2)^*\xi|\eta_1) = (\pi'(\eta_2^\flat)\xi|\eta_1)\\ &= (\pi(\xi)\eta_2^\flat|\eta_1) = (\eta_2^\flat|\pi(\xi^\#)\eta_1)\end{aligned}$$

$$= (\eta_2^b | \pi'(\eta_1)\xi^\#) = (\pi'(\eta_1)^* \eta_2^b | \xi^\#)$$

$$= (\pi'(\eta_1^b)\eta_2^b | \xi^\#) = (\eta_2^b \eta_1^b | \xi^\#) ,$$

so that $\eta_1\eta_2$ belongs to $\mathfrak{D}^b$ and

$$(3.4) \qquad (\eta_1\eta_2)^b = \eta_2^b\eta_1^b .$$

The π'-boundedness of $\eta_1\eta_2$ follows from the inequality:

$$\|\pi(\xi)\eta_1\eta_2\| = \|\pi(\xi)\pi'(\eta_2)\eta_1\|$$

$$= \|\pi'(\eta_2)\pi(\xi)\eta_1\| = \|\pi'(\eta_2)\pi'(\eta_1)\xi\|$$

$$\leq \|\pi'(\eta_2)\pi'(\eta_1)\|\|\xi\| .$$

Now, it is clear that π' preserves the involution and $\pi'(\eta_1\eta_2) = \pi'(\eta_2)\pi'(\eta_1)$. This completes the proof.

LEMMA 3.3. $\mathfrak{A}'$ and $(\mathfrak{A}')^2$ are both dense in the Hilbert space $\mathfrak{D}^b$; therefore they are dense in $\mathfrak{H}$.

<u>Proof</u>. Take and fix an $\eta \in \mathfrak{D}^b$. Put $a = \pi'(\eta)$ and let

$$a = uh = ku$$

be its polar decomposition, where $h = (a^*a)^{\frac{1}{2}}$ and $k = (aa^*)^{\frac{1}{2}}$. Then h and k are positive and affiliated with $\mathfrak{L}(\mathfrak{A})'$. Remark that, for every bounded continuous function f of a real variable,

$$(3.5) \qquad uf(h) = f(k)u \quad \text{and} \quad af(h) \supseteq f(k)a \quad .$$

Let f be a continuous function of a real variable with compact support. Then $h^2f(h)$ and $k^2f(k)$ both belong to $\mathfrak{L}(\mathfrak{A})'$. Put $\eta_1 = af(h)\eta^b$. Then we have, for each $\xi \in \mathfrak{A}$,

$$(\xi|\eta_1) = (\xi|af(h)\eta^b) = (\bar{f}(h)a^*\xi|\eta^b)$$

$$= (\bar{f}(h)\pi(\xi)\eta^b|\eta^b) = (\pi(\xi)\bar{f}(h)\eta^b|\eta^b)$$

$$= (\bar{f}(h)\eta^b|\pi(\xi^\#)\eta^b) = (\bar{f}(h)\eta^b|\pi'(\eta^b)\xi^\#)$$

$$= (a\bar{f}(h)\eta^b|\xi^\#) ;$$

hence η_1 belongs to $\mathfrak{D}^b$ and $\eta_1^b = a\bar{f}(h)\eta^b$. The calculation:

$$\pi(\xi)\,\eta_1 = \pi(\xi)af(h)\eta^b = af(h)\pi(\xi)\eta^b$$

$$= af(h)a^*\xi = k^2f(k)\xi, \ \xi \in \mathfrak{A} ,$$

shows that η_1 is π'-bounded and $k^2f(k) = \pi'(\eta_1)$. Hence we have, for $\xi \in \mathfrak{A}$,

$$(\xi|k^2\bar{f}(k)\eta) = (k^2f(k)\xi|\eta) = (\pi(\xi)\eta_1|\eta)$$

$$= (\eta_1|\pi(\xi^\#)\eta) = (\eta_1|\pi'(\eta)\xi^\#)$$

$$= (\pi'(\eta)^*\eta_1|\xi^\#) = (a^*af(h)\eta^b|\xi^\#)$$

$$= (h^2f(h)\eta^b|\xi^\#) ,$$

which implies that $k^2\bar{f}(k)\eta$ belongs to $\mathfrak{D}^b$ and

$$(k^2 f(k)\eta)^\flat = h^2 f(h)\eta^\flat \quad .$$

The π'-boundedness of $k^2 f(k)\eta$ follows from the calculation:

$$\|\pi(\xi)k^2 f(k)\eta\| = \|k^2 f(k)\pi(\xi)\eta\|$$

$$= \|k^2 f(k)a\xi\| = \|k^3 f(k)u\xi\|$$

$$\leq \|k^3 f(k)\|\|\xi\| \ .$$

Hence $k^2 f(k)\eta$ belongs to $\mathfrak{A}'$. Choose an increasing sequence $\{f_n\}$ of continuous positive functions of a real variable with compact support such that $\lim_{n\to\infty} t^2 f_n(t) = 1$ for $t > 0$. Then $\{k^2 f_n(k)\}$ converges strongly to the range projection of k, which is also the range projection of a. On the other hand, choosing a net $\{\xi_\alpha\}$ in $\mathfrak{A}$ such that $\{\pi(\xi_\alpha)\}$ converges strongly to the identity, we get

$$\eta = \lim \pi(\xi_\alpha)\eta = \lim \pi'(\eta)\xi_\alpha \ .$$

Hence η belongs to the closure of the range of a; so we have

$$\eta = \lim_{n\to\infty} k^2 f_n(k)\eta \ .$$

Furthermore, we have

$$\lim_{n\to\infty} (k^2 f_n(k)\eta)^\flat = \lim_{n\to\infty} h^2 f_n(h)\eta^\flat$$

$$= \eta^\flat \ ,$$

because $\eta^\flat$ belongs to the closure of the range of a^*. Therefore, η is well approximated by $\mathfrak{A}'$ in the metric of $\mathfrak{D}^\flat$. If we replace the above sequence $\{k^2 f_n(k)\}$ by $\{k^4 f_n(k)^2\}$, then we get the assertion for $(\mathfrak{A}')^2$.

In particular, $\pi'(\mathfrak{A}')$ is a non-degenerate *-algebra of operators on $\mathfrak{H}$ contained in $\mathfrak{L}(\mathfrak{A})'$.

LEMMA 3.4. If $\eta_1 \in \mathfrak{H}$ and $\eta_2 \in \mathfrak{H}$ satisfy the equation:

$$(3.6) \qquad (\xi_1^{\#}\xi_2|\eta_1) = (\eta_2|\xi_2^{\#}\xi_1), \ \xi_1,\xi_2 \in \mathfrak{A},$$

then η_1 and η_2 both belong to $\mathfrak{D}^\flat$, and $\eta_2 = \eta_1^\flat$.

<u>Proof</u>. By the remark just mentioned above, $\mathfrak{A}'$ contains a net $\{\eta_\alpha\}$ such that $\pi'(\eta_\alpha)$ converges strongly to the identity 1; hence we have

$$\xi = \lim \pi'(\eta_\alpha)\xi = \lim \pi(\xi)\eta_\alpha$$

for each $\xi \in \mathfrak{A}$, which means that each $\xi \in \mathfrak{A}$ is in the closure of the range of $\pi(\xi)$. Now, take and fix a ξ in $\mathfrak{A}$. Multiplying by a scalar if necessary, we may assume that $\|\pi(\xi)\| \leq 1$. Define a sequence $\{p_n\}$ of polynomials by:

$$(3.7) \qquad p_n(t) = 1 - (1 - t)^n .$$

Then the sequence $\{p_n(aa^*)\}$ converges strongly to the range projection of the bounded operator a if $\|a\| \leq 1$. Hence we have

$$(\xi|\eta_1) = \lim(p_n(\xi\xi^{\#})\xi|\eta_1)$$

$$= \lim(\eta_2|\xi^{\#}p_n(\xi\xi^{\#}))$$

$$= \lim(\eta_2 | p_n(\xi^\# \xi)\xi^\#)$$

$$= (\eta_2 | \xi^\#) ,$$

which implies our assertion.

Let $\mathfrak{D}^\#$ denote the definition domain of the adjoint involution of the involution: $\eta \mapsto \eta^b$. Namely, $\mathfrak{D}^\#$ is the set of all elements $\xi \in \mathfrak{H}$ such that there exists another element $\xi' \in \mathfrak{H}$ satisfying the equation:

$$(\xi | \eta) = (\eta^b | \xi')$$

for each $\eta \in \mathfrak{D}^b$. Putting $\xi^\# = \xi'$, we get a conjugate linear reflexive closed operator $\xi \to \xi^\#$ with domain $\mathfrak{D}^\#$ which is compatible with the original involution in $\mathfrak{A}$. We consider the inner product in $\mathfrak{D}^\#$ defined by:

$$(3.8) \qquad \langle \xi_1 | \xi_2 \rangle_\# = (\xi_1 | \xi_2) + (\xi_2^\# | \xi_1^\#), \ \xi_1, \xi_2 \in \mathfrak{D}^\# .$$

Then $\mathfrak{D}^\#$ is a Hilbert space with respect to the inner product (3.8). By Lemma 3.4, $\mathfrak{A}^2$ is dense in the Hilbert space $\mathfrak{D}^\#$. Furthermore, for $\xi \in \mathfrak{H}$ to be in $\mathfrak{D}^\#$, it is necessary and sufficient by Lemma 3.3 that there exists an element $\xi' \in \mathfrak{H}$ satisfying the equation:

$$(\xi | \eta_1 \eta_2^b) = (\eta_2 \eta_1^b | \xi')$$

for every $\eta_1, \eta_2 \in \mathfrak{A}'$.

LEMMA 3.5. If x is in $\mathfrak{L}(\mathfrak{A})'$ and η_1, η_2 are in $\mathfrak{A}'$, then $\pi'(\eta_1)x\eta_2$ belongs to $\mathfrak{A}'$, and

$$(\pi'(\eta_1)x\eta_2)^\flat = \pi'(\eta_2^\flat)x^*\eta_1^\flat \; ;$$

(3.9)

$$\pi'(\pi'(\eta_1)x\eta_2) = \pi'(\eta_1)x\pi'(\eta_2) \; .$$

<u>Proof</u>. For each $\xi \in \mathfrak{A}$, we have

$$\begin{aligned}(\xi|\pi'(\eta_1)x\eta_2) &= (\pi'(\eta_1^\flat)\xi|x\eta_2)\\ &= (\pi(\xi)\eta_1^\flat|x\eta_2)\\ &= (\eta_1^\flat|\pi(\xi^\#)x\eta_2)\\ &= (\eta_1^\flat|x\pi'(\eta_2)\xi^\#)\\ &= (\pi'(\eta_2^\flat)x^*\eta_1^\flat|\xi^\#) \; ;\end{aligned}$$

hence our assertion follows from Lemma 3.4.

Now we get the following commutation theorem for a generalized Hilbert algebra $\mathfrak{A}$.

THEOREM 3.1. $\mathfrak{L}(\mathfrak{A})'$ is generated by $\pi'(\mathfrak{A}')$.

<u>Proof</u>. By Lemma 3.5. $\pi'(\eta_1)x\pi'(\eta_2)$ belongs to $\pi'(\mathfrak{A}')$ for every $x \in \mathfrak{L}(\mathfrak{A})'$. Take a net $\{\eta_\alpha\}$ in $\mathfrak{A}'$ such that $\|\pi'(\eta_\alpha)\| \leq 1$ and $\{\pi'(\eta_\alpha)\}$ converges strongly to 1. Then we have

$$x = \lim \pi'(\eta_\alpha)x\pi'(\eta_\alpha) \; ,$$

which completes the proof.

For each $\xi \in \mathfrak{D}^\#$, define operators a and b by $a\eta = \pi'(\eta)\xi$

and $b\eta = \pi'(\eta)\xi^{\#}$ for each $\eta \in \mathfrak{A}'$. By similar arguments as for $\mathfrak{D}^{\flat}$, we can conclude that $a^* \supset b$ and $a \subset b^*$ and that a and b both commute with all operators in $\mathcal{L}(\mathfrak{A})'$, making use of Theorem 3.1. Let $\pi(\xi)$ denote the closure of a. Then we have $\pi(\xi)^* \supset \pi(\xi^{\#})$ and $\pi(\xi) \subset \pi(\xi^{\#})^*$; $\pi(\xi)$ and $\pi(\xi^{\#})$ are affiliated with $\mathcal{L}(\mathfrak{A})$.

DEFINITION 3.2. In the above situation, if $\pi(\xi)$ is bounded, then ξ is called π-bounded. In this case $\pi(\xi)$ belongs to $\mathcal{L}(\mathfrak{A})$.

Let $\mathfrak{A}''$ denote the set of all π-bounded elements. Define a product between elements $\xi \in \mathfrak{A}''$ and $\eta \in \mathfrak{H}$ by:

$$\xi\eta = \pi(\xi)\eta \ . \tag{3.10}$$

By the commutativity of $\pi(\xi)$ and $\pi'(\eta)$, $\xi \in \mathfrak{A}''$ and $\eta \in \mathfrak{A}'$, the products defined by (3.3) and (3.10) are compatible and satisfy the associativity law. By the same reason as for $\mathfrak{A}'$, $\mathfrak{A}''$ becomes an involutive algebra with the involution $\xi \mapsto \xi^{\#}$, which contains $\mathfrak{A}$ as a self-adjoint subalgebra. By Lemma 3.4, $\mathfrak{A}^2$ is dense in the Hilbert space $\mathfrak{D}^{\#}$.

Now, for the sake of convenience, we shall often denote the involution $\xi \mapsto \xi^{\#}$ and the adjoint involution $\eta \mapsto \eta^{\flat}$ as follows:

$$S\xi = \xi^{\#}, \ \xi \in \mathfrak{D}^{\#} \ ;$$

$$F\eta = \eta^{\flat}, \ \eta \in \mathfrak{D}^{\flat} \ .$$

4. The Commutation Theorem for Modular Hilbert Algebras

Suppose $\mathfrak{A}$ is a modular Hilbert algebra. Then by equalities (2.1) and (2.3), $\mathfrak{A}$ is a generalized Hilbert algebra, so that the results in §3 are all valid for $\mathfrak{A}$. Furthermore, $\mathfrak{A}$ is contained in $\mathfrak{A}'$. Hence the representation π' can be applied to $\mathfrak{A}$. By the equalities:

$$S\xi = J\Delta^{\frac{1}{2}}\xi = \Delta^{-\frac{1}{2}}J\xi \ ; \tag{4.1}$$

$$F\xi = J\Delta^{-\frac{1}{2}}\xi = \Delta^{\frac{1}{2}}J\xi, \ \xi \in \mathfrak{A} \ , \tag{4.2}$$

we have

$$\|\xi^{\#}\| = \|\Delta^{\frac{1}{2}}\xi\| \quad \text{and} \quad \|\xi^{\flat}\| = \|\Delta^{-\frac{1}{2}}\xi\| \ .$$

Since $\mathfrak{D}^{\#}$ is the closure of $\mathfrak{A}$ with respect to the $\| \ \|_{\#}$-norm and $\mathfrak{A}$ is dense in the Hilbert space $\mathfrak{D}(\Delta^{\frac{1}{2}})$, $\mathfrak{D}^{\#}$ and $\mathfrak{D}(\Delta^{\frac{1}{2}})$ are identical as Hilbert spaces. Therefore, equality (4.1) implies the following:

$$S = \Delta^{-\frac{1}{2}}J = J\Delta^{\frac{1}{2}} \ , \tag{4.3}$$

which is nothing but the polar decomposition of the conjugate linear closed operator S. Since S, F and J leave $\mathfrak{A}$ invariant, we have, for each $\eta \in \mathfrak{D}^{\flat}$ and $\xi \in \mathfrak{A}$,

$$(\xi | JF\eta) = (F\eta | J\xi) = (SJ\xi | \eta)$$

$$= (\Delta^{-\frac{1}{2}}\xi | \eta) \ .$$

Since $\mathfrak{A}$ is dense in the Hilbert space $\mathfrak{D}(\Delta^{-\frac{1}{2}})$, η belongs to

$\mathfrak{D}(\Delta^{-\frac{1}{2}})$ and $\Delta^{-\frac{1}{2}}\eta = JF\eta$. Therefore $\mathfrak{D}^{b}$ and $\mathfrak{D}(\Delta^{-\frac{1}{2}})$ are identical as Hilbert spaces, and we have

$$(4.4) \qquad F = J\Delta^{-\frac{1}{2}} = \Delta^{\frac{1}{2}}J .$$

Thus J maps $\mathfrak{D}^{\#}$ isometrically onto $\mathfrak{D}^{b}$. Since $J\mathfrak{A}^2 = \mathfrak{A}^2$ and $\mathfrak{A}^2$ is dense in $\mathfrak{D}^{\#}$ by Lemma 3.4, $\mathfrak{A}^2$ is dense in $\mathfrak{D}^{b}$. Therefore, we get the following.

LEMMA 4.1. If $\mathfrak{A}$ is a modular Hilbert algebra, then for $\xi \in \mathfrak{H}$ to be in $\mathfrak{D}^{\#}$ it is necessary and sufficient that there is an element $\xi' \in \mathfrak{H}$ such that

$$(\xi|\eta_1 \cdot \eta_2^{b}) = (\eta_2 \cdot \eta_1^{b}|\xi')$$

for every $\eta_1, \eta_2 \in \mathfrak{A}$. If this is the case, then $\xi' = \xi^{\#}$.

LEMMA 4.2. If $\mathfrak{A}$ is a modular Hilbert algebra, then for $\xi \in \mathfrak{D}^{\#}$ to be π-bounded, it is necessary and sufficient that there is a constant $\gamma > 0$ such that

$$\|\pi'(\eta)\xi\| \leq \gamma\|\eta\|$$

for every $\eta \in \mathfrak{A}$. If this is the case, then $\|\pi(\xi)\| \leq \gamma$.

Proof. The necessity is trivial. Take and fix a $\xi \in \mathfrak{D}^{\#}$ and an $\eta \in \mathfrak{A}'$. We can find, by the density of $\mathfrak{A}$ in $\mathfrak{D}^{b}$, a sequence $\{\eta_n\}$ in $\mathfrak{A}$ such that $\lim \eta_n = \eta$. Then we have

$$\|\pi(\xi)\eta_n - \pi(\xi)\eta_m\| \leq \gamma\|\eta_n - \eta_m\| \to 0$$

as $n,m \to \infty$, so that η belongs to the definition domain of $\pi(\xi)$ by the closedness of $\pi(\xi)$, and $\pi(\xi)\eta = \lim_{n\to\infty} \pi(\xi)\eta_n$; hence we have

$$\|\pi(\xi)\eta\| = \lim \|\pi(\xi)\eta_n\| \leq \gamma \lim \|\eta_n\|$$

$$= \gamma\|\eta\| ,$$

which means the π-boundedness of ξ.

Combining these lemmas and making use of dual arguments of the proof of Theorem 3.1, we get the following commutation theorem.

THEOREM 4.1. If $\mathfrak{A}$ is a modular Hilbert algebra, then $\pi'(\mathfrak{A})$ generates the commutant $\mathcal{L}(\mathfrak{A})'$ of the left algebra, and the unitary involution J carries the von Neumann algebra $\mathcal{L}(\mathfrak{A})$ onto the commutant $\mathcal{L}(\mathfrak{A})'$, that is,

$$J\mathcal{L}(\mathfrak{A})J = \mathcal{L}(\mathfrak{A})', \quad J\mathcal{L}(\mathfrak{A})'J = \mathcal{L}(\mathfrak{A}) .$$

<u>Proof</u>. Since $\mathfrak{A} \subset \mathfrak{A}'$, $\pi(\mathfrak{A}'')$ is contained in $\pi'(\mathfrak{A})'$. We claim that $\pi(\mathfrak{A}'')$ generates $\pi'(\mathfrak{A})'$. Take and fix an x in $\pi'(\mathfrak{A})'$. For any $\xi_1, \xi_2 \in \mathfrak{A}''$ and $\eta_1, \eta_2 \in \mathfrak{A}$, we have, as in the proof of Lemma 3.5,

$$(\eta_1\eta_2^\flat | \pi(\xi_1)x\xi_2) = (\pi(\xi_2^\#)x^*\xi_1 | \eta_2\eta_1^\flat)$$

and $\pi'(\eta_1)x\pi(\xi_2)$ is in $\pi(\mathfrak{A}'')$. Take a net $\{\xi_\alpha\}$ in $\mathfrak{A}''$ such that $\|\pi(\xi_\alpha)\| \leq 1$ and $\pi(\xi_\alpha)$ converges strongly to 1. Then we have

$$x = \lim \pi(\xi_\alpha) x \pi(\xi_\alpha) \qquad \text{(strongly)} .$$

Hence $\pi(\mathfrak{A}'')$ is strongly dense in $\pi'(\mathfrak{A})'$. Therefore we have $\pi'(\mathfrak{A})' = \mathcal{L}(\mathfrak{A})$ and $\mathcal{L}(\mathfrak{A})' = \pi'(\mathfrak{A})''$. Since $J\mathfrak{A} = \mathfrak{A}$ and $J\pi(\xi)J = \pi'(J\xi)$ for $\xi \in \mathfrak{A}$, we get the second assertion.

Remark. $J\mathfrak{A}'' = \mathfrak{A}'$ and $J\mathfrak{A}' = \mathfrak{A}''$.

5. Self-Adjoint Subalgebras of Generalized Hilbert Algebras

LEMMA 5.1. Every self-adjoint subalgebra of a generalized Hilbert algebra (i.e. a subalgebra invariant under the involution: $\xi \to \xi^{\#}$) is a generalized Hilbert algebra.

Proof. Let $\mathfrak{A}$ be a generalized Hilbert algebra and $\mathfrak{B}$ be a self-adjoint subalgebra. Then postulates (I), (II) and (IX) for the generalized Hilbert algebra hold for $\mathfrak{B}$. If ξ is in $\mathfrak{B}$ and $\|\pi(\xi)\| \leq 1$, then $\xi = \lim_{n\to\infty} p_n(\xi\xi^{\#})\xi$, where $\{p_n\}$ is the sequence of polynomials defined by equality (3.7), so that $\mathfrak{B}$ satisfies postulate (III). Therefore $\mathfrak{B}$ is a generalized Hilbert algebra. This completes the proof.

It is not clear whether the corresponding assertion remains true for modular Hilbert algebras, because of postulate (VIII).

Throughout the rest of this section, suppose a self-adjoint subalgebra $\mathfrak{B}$ of a generalized Hilbert algebra $\mathfrak{A}$ is dense in $\mathfrak{A}$. In this case, the associated von Neumann algebras $\mathcal{L}(\mathfrak{A})$ and $\mathcal{L}(\mathfrak{B})$ both act on the same Hilbert space $\mathfrak{H}$. Of course, $\mathcal{L}(\mathfrak{A}) \supset \mathcal{L}(\mathfrak{B})$. By definition we have $\mathfrak{B}' \supset \mathfrak{A}'$ and $\mathfrak{B}'' \subset \mathfrak{A}''$.

LEMMA 5.2. $\mathfrak{B}'' = \mathfrak{A}''$ if and only if $\mathfrak{B}$ is dense in $\mathfrak{A}$ with respect to the norm topology in the Hilbert space $\mathfrak{H}^{\#}$, constructed from $\mathfrak{A}$.

Proof. Let $\mathfrak{H}^{\#}(\mathfrak{A})$, $\mathfrak{H}^{b}(\mathfrak{A})$, $\mathfrak{H}^{\#}(\mathfrak{B})$ and $\mathfrak{H}^{b}(\mathfrak{B})$ denote the Hilbert spaces corresponding naturally to $\mathfrak{A}$ and $\mathfrak{B}$ respectively. Suppose $\mathfrak{B}$ is sense in $\mathfrak{A}$. Then the involution $\xi \to \xi^{\#}$ in $\mathfrak{H}^{\#}(\mathfrak{A})$ is the closure of the involution in $\mathfrak{B}$ as a closed operator, so that the involutions of $\mathfrak{A}$ and of $\mathfrak{B}$ have the same adjoint involution $\eta \to \eta^{b}$. Hence we have

$$\mathfrak{H}^{b}(\mathfrak{A}) = \mathfrak{H}^{b}(\mathfrak{B}) .$$

Take an $\eta \in \mathfrak{B}'$. Since η is in $\mathfrak{H}^{b}(\mathfrak{A})$, there exists a closed operator a affiliated with $\mathfrak{L}(\mathfrak{A})'$ such that

$$a\xi = \pi(\xi)\eta, \ \xi \in \mathfrak{A} .$$

Since η is in $\mathfrak{B}'$, there is a constant $\gamma > 0$ with $\|a\xi\| \leq \gamma\|\xi\|$ for each $\xi \in \mathfrak{B}$. Take a ξ in $\mathfrak{A}$. Then we can find a sequence $\{\xi_n\}$ in $\mathfrak{B}$ with $\lim \xi_n = \xi$. By the inequality:

$$\|a\xi_n - a\xi_m\| \leq \gamma\|\xi_n - \xi_m\| ,$$

$\{a\xi_n\}$ is a Cauchy sequence, so that by the closedness of a ξ belongs to the domain of a and $a\xi = \lim a\xi_n$; hence

$$\|a\xi\| = \lim\|a\xi_n\| \leq \gamma \lim\|\xi_n\| = \gamma\|\xi\| .$$

Hence a is bounded and in $\mathfrak{L}(\mathfrak{A})'$, which means that η is in $\mathfrak{A}'$; therefore $\mathfrak{B}' = \mathfrak{A}'$; equivalently $\mathfrak{B}'' = \mathfrak{A}''$.

The converse assertion follows from the fact that $\mathfrak{B}$ is dense in $\mathfrak{B}''$ with respect to the $\|\cdot\|_{\#}$-norm topology.

DEFINITION 5.1. Let $\mathfrak{A}$ and $\mathfrak{B}$ be two generalized Hilbert algebras. If $\mathfrak{B}$ is isometrically imbedded in $\mathfrak{A}''$ as a self-adjoint subalgebra and $\mathfrak{B}'' = \mathfrak{A}''$, then $\mathfrak{A}$ and $\mathfrak{B}$ are called equivalent. Of course, if they are equivalent, then $\mathfrak{L}(\mathfrak{A})$ and $\mathfrak{L}(\mathfrak{B})$ are isomorphic. If $\mathfrak{A} = \mathfrak{A}''$, then $\mathfrak{A}$ is called <u>achieved</u>. Clearly every generalized Hilbert algebra is equivalent to an achieved generalized Hilbert algebra.

6. The Spectral Algebra

Suppose $\mathfrak{A}$ is an achieved generalized Hilbert algebra. i.e. $\mathfrak{A}'' = \mathfrak{A}$. Take and fix a ξ in $\mathfrak{D}^{\#}$. Let

(6.1) $$\pi(\xi) = uh = ku$$

be the left and right polar decompositions of the closed operator $\pi(\xi)$, which will be denoted by a, $h = (\pi(\xi)^{*}\pi(\xi))^{\frac{1}{2}}$ and $k = (\pi(\xi)\pi(\xi)^{*})^{\frac{1}{2}}$. Let $\mathcal{K}$ be the space of all continuous complex valued functions of a real variable with compact support contained in the open interval $(0,\infty)$. Let f_0 denote the function defined by $f_0(\lambda) = \lambda$. It is clear that

$$f_0\mathcal{K} = \mathcal{K} \quad \text{and} \quad f_0^{-1}\mathcal{K} = \mathcal{K} .$$

For each $f \in \mathcal{K}$, $f(k)$ and $f(h)$ are both bounded and contained in $\mathfrak{L}(\mathfrak{A})$. As in (3.5), we have

(6.2) $$f(k)a \subseteq af(h) \quad \text{and} \quad a^{*}f(k) \supseteq f(h)a^{*}$$

for every $f \in \mathcal{K}$ and all these operators are bounded; so we may regard (6.2) as equalities. For each $f \in \mathcal{K}$, put

$$\eta(f) = (f_0^{-1}f)(k)u\xi^{\#} . \tag{6.3}$$

LEMMA 6.1. The $\eta(f)$ with $f \in \mathcal{K}$ form an abelian self-adjoint subalgebra of $\mathfrak{A}$, and we have

$$\pi(\eta(f)) = f(k) . \tag{6.4}$$

Proof. Put

$$g(\lambda) = f_0(\lambda)^{-1}f(\lambda) = \frac{1}{\lambda} f(\lambda), \ 0 < \lambda < \infty .$$

For each $\zeta_1, \zeta_2 \in \mathfrak{A}'$, we have

$$\begin{aligned}
(\eta(f)|\zeta_1\zeta_2^{\flat}) &= (\pi'(\zeta_2)g(k)u\xi^{\#}|\zeta_1) \\
&= (g(k)u\pi'(\zeta_2)\xi^{\#}|\zeta_1) \\
&= (g(k)ua^{*}\zeta_2|\zeta_1) \\
&= (g(k)k\zeta_2|\zeta_1) = (\zeta_2|k\bar{g}(k)\zeta_1) \\
&= (\zeta_2|ua^{*}\bar{g}(k)\zeta_1) = (\zeta_2|u\bar{g}(h)a^{*}\zeta_1) \\
&= (\zeta_2|u\bar{g}(h)\pi'(\zeta_1)\xi^{\#}) = (\pi'(\zeta_1)^{*}\zeta_2|u\bar{g}(h)\xi^{\#}) \\
&= (\zeta_2\zeta_1^{\flat}|\bar{g}(k)u\xi^{\#}) = (\zeta_2\zeta_1^{\flat}|\eta(\bar{f})) .
\end{aligned}$$

Therefore, by Lemma 3.4 $\eta(f)$ belongs to $\mathcal{D}^{\#}$ and

$$\eta(f)^{\#} = \eta(\bar{f}), \ f \in \mathcal{K} . \tag{6.5}$$

Also the above calculation assures equality (6.4). Hence we have

$$(6.6)\qquad \eta(f\bar{g}) = f(k)\eta(\bar{g}) = \eta(f)\eta(g)^{\#},\ f,\ g \in \mathcal{K},$$

which implies our first assertion.

DEFINITION 6.1. The algebra of all $\eta(f)$, $f \in \mathcal{K}$, is called the spectral algebra of ξ and is denoted by $\mathfrak{A}_0(\xi)$.

LEMMA 6.2. There exists a positive Radon measure μ on the open interval $(0,\infty)$ such that the map: $\eta \in \mathcal{K} \to \eta(f)$ is extended to an isometry $\tilde{\eta}$ of $L^2(\mu)$ onto the closure $[\mathfrak{A}_0(\xi)]$ of $\mathfrak{A}_0(\xi)$ in $\mathfrak{H}$ with the following property:

$$(6.7)\qquad \tilde{\eta}(fg) = f(k)\tilde{\eta}(g)$$

for every $f \in L^\infty(\mu)$ and $g \in L^2(\mu)$.

Proof. Put

$$\mu\left(\sum_{i=1}^{n} f_i\overline{g_i}\right) = \sum_{i=1}^{n} (\eta(f_i)| \qquad))$$

for $f_i, g_i \in \mathcal{K}$, $i = 1,2,\ldots,n$. Suppose $\qquad f_i\bar{g}_i = 0$. Take an element $h \in \mathcal{K}$ such that $h(\lambda) = 1$ if $\qquad \bigcup_{i=1}^{n}$ supp. g_i. Then we have

$$\sum_{i=1}^{n} (\eta(f_i)|\eta(g_i)) = \sum_{i=1}^{n} (\eta(f_i)|\eta(g_ih))$$

$$= \sum_{i=1}^{n} (\eta(f_i)|g_i(k)\eta(h)) \qquad \text{by (6.6)}$$

$$= \sum_{i=1}^{n} (\bar{g}_i(k)\eta(f_i)|\eta(h))$$

$$= \Big(\sum_{i=1}^{n} \eta(f_i\bar{g}_i)|\eta(h)\Big) = \Big(\eta \Big(\sum_{i=1}^{n} f_i\bar{g}_i\Big)\Big|\eta(h)\Big)$$

$$= 0 \ .$$

Therefore, μ is a linear functional on $\mathcal{K}$. Furthermore, μ is positive, because

$$(6.8) \qquad \mu(|f|^2) = \|\eta(f)\|^2 \geq 0 \ ;$$

hence μ is a measure on $(0,\infty)$. By equality (6.8), η is extended to an isometry $\tilde{\eta}$ of $L^2(\mu)$ into $\mathfrak{H}$. Since $\mathcal{K}$ is dense in $L^2(\mu)$ and $\eta(\mathcal{K}) = \mathfrak{A}_0(\xi)$, the range of $\tilde{\eta}$ is the closure $[\mathfrak{A}_0(\xi)]$ of $\mathfrak{A}_0(\xi)$. By (6.6), $\tilde{\eta}$ induces a spatial isomorphism of the operator algebra $\mathcal{K}$, acting on $L^2(\mu)$ as multiplication operators, onto the operator algebra $\pi(\mathfrak{A}_0(\xi))$ acting on $[\mathfrak{A}_0(\xi)]$. Therefore $\tilde{\eta}$ induces a spacial isomorphism of the operator algebra $L^\infty(\mu)$ acting on $L^2(\mu)$ onto the weak closure of $\pi(\mathfrak{A}_0(\xi))$ acting on $[\mathfrak{A}_0(\xi)]$, which implies equality (6.7). This completes the proof.

LEMMA 6.3. The spectral algebra $\mathfrak{A}_0(\xi)$ has the following properties:

(i) $\mathfrak{A}_0(\xi)\xi$ is contained in $\mathfrak{A}$ and

$$(6.9) \qquad ((\eta^\#\eta\xi)^\#|\xi^\#) \geq 0$$

for every $\eta \in \mathfrak{A}_0(\xi)$;

(ii) ξ belongs to $\mathfrak{A}$ if and only if there is a constant $\gamma > 0$ with

$$(6.10) \qquad ((\eta^{\#}\eta\xi)^{\#}|\xi^{\#}) \leq \gamma^2\|\eta\|^2, \ \eta \in \mathfrak{A}_0(\xi) \ ;$$

if this is the case, then $\|\pi(\xi)\| \leq \gamma$.

Proof. Take an $f \in \mathcal{K}$. By (6.4), we have $\eta(f)\xi = f(k)\xi$. For each $\zeta_1, \zeta_2 \in \mathfrak{A}'$, we have

$$\begin{aligned}(\eta(f)\xi|\zeta_1\zeta_2^{\flat}) &= (f(k)\xi|\pi'(\zeta_2)^*\zeta_1)\\ &= (\pi'(\zeta_2)f(k)\xi|\zeta_1) = (f(k)\pi'(\zeta_2)\xi|\zeta_1)\\ &= (f(k)a\zeta_2|\zeta_1) = (f(k)ku\zeta_2|\zeta_1)\\ &= (\zeta_2|u^*k\bar{f}(k)\zeta_1) = (\zeta_2|\bar{f}(h)a^*\zeta_1)\\ &= (\zeta_2|\bar{f}(h)\pi'(\zeta_1)\xi^{\#}) = (\zeta_2\zeta_1|\bar{f}(h)\xi^{\#}) ,\end{aligned}$$

so that

$$(6.11) \qquad (\eta(f)\xi)^{\#} = \bar{f}(h)\xi^{\#} \quad \text{and} \quad \pi(\eta(f)\xi) = f(k)a$$

for each $f \in \mathcal{K}$. Hence $\eta(f)\xi$ belongs to $\mathfrak{A}$ and also

$$\begin{aligned}((\eta(f)^{\#}\eta(f)\xi)^{\#}|\xi^{\#}) &= ((\eta(\bar{f}\cdot f)\xi)^{\#}|\xi^{\#})\\ &= (\bar{f}(h)f(h)\xi^{\#}|\xi^{\#}) = \|f(h)\xi^{\#}\|^2 \geq 0 .\end{aligned}$$

Suppose there is a constant γ satisfying inequality (6.10). Then we have

$$\|k\eta(f)\|^2 = \|f(k)u\xi^{\#}\|^2 = \|uf(h)\xi^{\#}\|^2$$
$$= \|f(h)\xi^{\#}\|^2 \leq \gamma^2\|h(f)\|^2 ,$$

which means that

$$\int_0^\infty |\lambda f(\lambda)|^2 d\mu(\lambda) \leq \gamma^2 \int_0^\infty |f(\lambda)|^2 d\mu(\lambda)$$

for every $f \in \mathfrak{K}$. Therefore, the measure μ is supported by the closed interval $[0,\gamma]$; and then $\eta(f) = 0$ if $\operatorname{supp} f$ is contained in the open interval (γ,∞), so that $\pi(\eta(f)) = f(k) = 0$. Therefore, the spectrum of k is contained in the interval $[0,\gamma]$ which means $\|k\| \leq \gamma$; hence ξ is π-bounded.

It is not difficult to show the converse assertion. Since it will not be used in the sequel, we omit the proof.

As the dual assertion of Lemma 6.3, we get the following:

LEMMA 6.3'. For each $\xi \in \mathfrak{D}^b$, there exists an abelian self-adjoint subalgebra $\mathfrak{A}_0'(\xi)$ of $\mathfrak{A}'$ with the following properties:

(i) $\xi\mathfrak{A}_0'(\xi)$ is contained in $\mathfrak{A}'$ and

$$((\xi\eta\eta^b)^b|\xi^b) \geq 0 \tag{6.9}$$

for every $\eta \in \mathfrak{A}_0'(\xi)$;

(ii) ξ belongs to $\mathfrak{A}'$ if and only if there is a constant $\gamma > 0$ such that

$$((\xi\eta\eta^b)^b|\xi^b) \leq \gamma^2\|h\|^2, \ \eta \in \mathfrak{A}_0'(\xi) ; \tag{6.10}$$

if this is the case, then $\|\pi'(\xi)\| \leq \gamma$.

Proof. Considering the reversed product in $\mathfrak{A}'$ such as $\xi \cdot \eta = \eta\xi$, we get an achieved generalized Hilbert algebra $\mathfrak{A}'$. Then the roles of $\mathfrak{A}$ and $\mathfrak{A}'$ are completely interchanged in this new situation. Our assertion is nothing but the interpretation of the former result, Lemma 6.3, in the new situation. This completes the proof.

7. The Modular Operator Δ

Now, we are in the position to introduce the modular operator Δ of a generalized Hilbert algebra $\mathfrak{A}$, which will play a key role throughout the theory.

Suppose $\mathfrak{A}$ is an achieved generalized Hilbert algebra in the sequel. As was shown in §3, $\mathfrak{D}^{\#}$ and $\mathfrak{D}^{\flat}$ are both Hilbert spaces with the inner products $\langle\cdot|\cdot\rangle_{\#}$ and $\langle\cdot|\cdot\rangle_{\flat}$ respectively. The original inner product $(\cdot|\cdot)$ is bounded by the new one $\langle\cdot|\cdot\rangle_{\#}$ in $\mathfrak{D}^{\#}$, so that there exists a positive operator H on $\mathfrak{D}^{\#}$ such that

$$\langle H\xi|\eta\rangle_{\#} = (\xi|\eta),\ \xi, \eta \in \mathfrak{D}^{\#} ,$$

(7.1)

$$0 \leq H \leq 1 .$$

Recalling that S is an isometric involution in $\mathfrak{D}^{\#}$, we have

$$\begin{aligned}\langle SHS\xi|\eta\rangle_{\#} &= \langle S\eta|HS\xi\rangle_{\#} = \langle HS\eta|S\xi\rangle_{\#}\\ &= (S\eta|S\xi) = \langle\xi|\eta\rangle_{\#} - (\xi|\eta)\\ &= \langle\xi|\eta\rangle_{\#} - \langle H\xi|\eta\rangle_{\#} = \langle(1 - H)\xi|\eta\rangle_{\#} ,\end{aligned}$$

which means that

$$SHS = (1 - H) . \tag{7.2}$$

Let i denote the injection of $\mathfrak{D}^{\#}$ into $\mathfrak{H}$. Then H is nothing but $i^* i$. Let

$$i = VH^{\frac{1}{2}} = K^{\frac{1}{2}}V \tag{7.3}$$

be the polar decomposition of i, where $K = ii^*$. Then we have, for every continuous function f defined on the closed interval $[0,1]$,

$$if(H) = f(K)i . \tag{7.4}$$

Therefore, $f(K)$ is an extension of $f(H)$ on $\mathfrak{H}$. As a particular case, $K^{\frac{1}{2}}$ is also an extension of $H^{\frac{1}{2}}$ on $\mathfrak{H}$. Hence we have, for $\xi \in \mathfrak{D}^{\#}$,

$$\|\xi\|^2 = \|H^{\frac{1}{2}}\xi\|_{\#}^2 = \|K^{\frac{1}{2}}\xi\|_{\#}^2 , \tag{7.5}$$

so that $K^{\frac{1}{2}}$ maps the dense set $\mathfrak{D}^{\#}$ of $\mathfrak{H}$ isometrically into the Hilbert space $\mathfrak{D}^{\#}$. Since the range of $K^{\frac{1}{2}}$ contains that of $H^{\frac{1}{2}}$, the range of $K^{\frac{1}{2}}$ is dense in the Hilbert space of $\mathfrak{D}^{\#}$, so that $K^{\frac{1}{2}}$ is an isometry of $\mathfrak{H}$ onto the Hilbert space $\mathfrak{D}^{\#}$, that is, $\mathfrak{D}^{\#}$ is the range of the positive bounded operator $K^{\frac{1}{2}}$. Define a positive self-adjoint operator Δ (not necessarily bounded) by:

$$\Delta = K^{-1}(1 - K) . \tag{7.6}$$

Then the domain $\mathfrak{D}(\Delta^{\frac{1}{2}})$ of $\Delta^{\frac{1}{2}}$ coincides with the range of $K^{\frac{1}{2}}$

as a set, so that $\mathfrak{D}(\Delta^{\frac{1}{2}})$ and $\mathfrak{D}^{\#}$ are identical as subsets of $\mathfrak{H}$. By equalities (7.4) and (7.5) we have, for each $\xi, \eta \in \mathfrak{D}(\Delta^{\frac{1}{2}})$,

$$\begin{aligned}
\langle \xi | \eta \rangle_{\Delta^{\frac{1}{2}}} &= (\xi|\eta) + (\Delta^{\frac{1}{2}}\xi | \Delta^{\frac{1}{2}}\eta) \\
&= (\xi|\eta) + (K^{-\frac{1}{2}}(1 - K)^{\frac{1}{2}}\xi | K^{-\frac{1}{2}}(1 - K)^{\frac{1}{2}}\eta) \\
&= (\xi|\eta) + \langle (1 - K)^{\frac{1}{2}}\xi | (1 - K)^{\frac{1}{2}}\eta \rangle_{\#} \\
&= \langle H\xi|\eta\rangle_{\#} + \langle (1 - H)^{\frac{1}{2}}\xi | (1 - H)^{\frac{1}{2}}\eta \rangle_{\#} \\
&= \langle \xi|\eta\rangle_{\#}
\end{aligned}$$

Therefore, $\mathfrak{D}(\Delta^{\frac{1}{2}})$ and $\mathfrak{D}^{\#}$ are identical not only as sets but also as Hilbert spaces.

Put $J = VSV^*$. Then J is a reflexive conjugate linear isometry of $\mathfrak{H}$. By equality (7.2), we have

$$JKJ = (1 - K) ,$$

so that we have

$$(7.7) \qquad J\Delta J = \Delta^{-1} .$$

Furthermore, for every measurable function f defined on the open interval $(0,\infty)$ we have

$$(7.7') \qquad Jf(\Delta)J = \bar{f}(\Delta^{-1}) .$$

Since $V = K^{-\frac{1}{2}}$ and $V^* = K^{\frac{1}{2}}$ by equality (7.3) and the above arguments, we have, for $\xi \in \mathfrak{D}^{\#}$,

$$S\xi = V^*JV\xi = K^{\frac{1}{2}}JK^{-\frac{1}{2}}\xi$$

$$= J(1 - K)^{\frac{1}{2}}K^{-\frac{1}{2}}\xi$$

$$= J\Delta^{\frac{1}{2}}\xi \ .$$

Hence we get the polar decomposition

(7.8) $$S = J\Delta^{\frac{1}{2}} = \Delta^{-\frac{1}{2}}J$$

of the involution S. For $\xi \in \mathfrak{H}$ to be in $\mathfrak{D}^{\flat}$, it is necessary and sufficient that there exists a vector $\xi^{\flat} \in \mathfrak{H}$ such that

(7.9) $$(S\eta | \xi) = (\xi^{\flat} | \eta), \ \eta \in \mathfrak{D}^{\#} .$$

Noticing that $J\mathfrak{D}^{\#} = \mathfrak{D}(\Delta^{-\frac{1}{2}})$ and $J\mathfrak{D}(\Delta^{-\frac{1}{2}}) = \mathfrak{D}^{\#}$, we have, for each $\eta \in \mathfrak{D}(\Delta^{-\frac{1}{2}})$,

$$(J\xi^{\flat} | \eta) = (J\eta | \xi^{\flat}) = (\xi | SJ\eta) = (\xi | \Delta^{-\frac{1}{2}}\eta) \ .$$

Hence ξ belongs to $\mathfrak{D}(\Delta^{-\frac{1}{2}})$ and $J\xi^{\flat} = \Delta^{-\frac{1}{2}}\xi$; so that $\xi^{\flat} = J\Delta^{-\frac{1}{2}}\xi$. Conversely, if ξ belongs to $\mathfrak{D}(\Delta^{-\frac{1}{2}})$, then $\xi^{\flat} = J\Delta^{-\frac{1}{2}}\xi$ satisfies equation (7.9). Therefore, we get

$$\mathfrak{D}^{\flat} = \mathfrak{D}(\Delta^{-\frac{1}{2}}) \ \text{ and } \ \xi^{\flat} = J\Delta^{-\frac{1}{2}}\xi, \ \xi \in \mathfrak{D}^{\flat} .$$

Since J is an isometry, $\mathfrak{D}^{\flat}$ and $\mathfrak{D}(\Delta^{-\frac{1}{2}})$ are identical as Hilbert spaces. Thus we get the following:

THEOREM 7.1. There exist an isometric involution J and a positive self-adjoint operator Δ such that

(i) $\mathfrak{D}^{\#} = \mathfrak{D}(\Delta^{\frac{1}{2}})$ and $\mathfrak{D}^{\flat} = \mathfrak{D}(\Delta^{-\frac{1}{2}})$;

(ii) $J\Delta J = \Delta^{-1}$ and $Jf(\Delta)J = \bar{f}(\Delta^{-1})$

for every measurable function f defined on the open interval $(0,\infty)$;

(iii) $S = J\Delta^{\frac{1}{2}} = \Delta^{-\frac{1}{2}}J,$

$F = J\Delta^{-\frac{1}{2}} = \Delta^{\frac{1}{2}}J;$

(iv) $\Delta = FS$ and $\Delta^{-1} = SF$.

The operator Δ is called the <u>modular</u> operator of the generalized Hilbert algebra $\mathfrak{A}$ and J the <u>unitary involution</u> of $\mathfrak{A}$. J induces a transpose operation: $x \in \mathcal{B}(\mathcal{H}) \to x^T \in \mathcal{B}(\mathcal{H})$, which is an anti-automorphism of $\mathcal{B}(\mathcal{H})$, as follows:

$$x^T = Jx^*J, \; x \in \mathcal{B}(\mathcal{H}) \,, \tag{7.10}$$

where $\mathcal{B}(\mathcal{H})$ denotes the algebra of all bounded operators on $\mathcal{H}$.

8. The Resolvent of the Modular Operator Δ

Keep the assumptions and the notations in §7. Let $\tilde{C}$ denote the Riemann sphere $C \cup \{\infty\}$ and $[0,\infty]$ denote the extended positive half line $\{z \in \tilde{C} : 0 \leq z \leq \infty\}$. Let $A[0,\infty]$ denote the space of all functions f which are analytic in a neighborhood U_f of $[0,\infty]$ and vanish at infinity. We shall study $f(\Delta)$, $f \in A[0,\infty]$. For each $\omega \in \tilde{C} - [0,\infty]$, put

$$\gamma(\omega) = (2|\omega| - \omega - \bar{\omega})^{-\frac{1}{2}} \,;$$

(8.1)

$$R(\omega) = (\Delta - \omega)^{-1} \,.$$

Then by Theorem 7.1, we have

$$R(\omega)^T = (\Delta^{-1} - \omega)^{-1} . \tag{8.2}$$

LEMMA 8.1. If $\omega \in \tilde{C} - [0,\infty]$, we have

$$R(\omega)\mathfrak{A}' \subset \mathfrak{A} ,$$

$$\|\pi(R(\omega')\xi')\| \leq \gamma(\omega)\|\pi'(\xi')\|, \ \xi' \in \mathfrak{A} ;$$

$$R(\omega)^T\mathfrak{A} \subset \mathfrak{A}' ;$$

$$\|\pi'(R(\omega)^T\xi)\| \leq \gamma(\omega)\|\pi(\xi)\|, \ \xi \in \mathfrak{A} .$$

Proof. Take and fix a $\xi' \in \mathfrak{A}'$. Put

$$\xi = R(\omega)\xi' = (\Delta - \omega)^{-1}\xi' .$$

Then ξ is in $\mathcal{D}(\Delta)$; so it is in $\mathcal{D}^{\#}$. For each η in the spectral algebra $\mathfrak{A}_0(\xi)$ of ξ, we have

$$\begin{aligned}
(2|\omega| - \omega - \bar{\omega})((\eta^{\#}\eta\xi)^{\#}|\xi^{\#}) &= (2|\omega| - \omega - \bar{\omega})(\Delta\xi|\eta^{\#}\eta\xi) \\
&= (2|\omega| - \omega - \bar{\omega})(\eta(\Delta\xi)|\eta\xi) \\
&\leq 2|\omega|\,\|\eta(\Delta\xi)\|\,\|\eta\xi\| - 2 \operatorname{Re} \omega(\eta(\Delta\xi)|\eta\xi) \\
&\leq \|\eta(\Delta\xi)\|^2 + |\omega|^2\|\eta\xi\|^2 - 2 \operatorname{Re} \omega(\eta(\Delta\xi)|\eta\xi) \\
&= \|\eta((\Delta - \omega)\xi)\|^2 = \|\eta\xi'\|^2 \leq \|\pi'(\xi')\|^2\|\eta\|^2 .
\end{aligned}$$

Hence we get

$$((\eta^{\#}\eta\xi)^{\#}|\xi^{\#}) \leq \gamma(\omega)^2\|\pi'(\xi')\|^2\|\eta\|^2, \ \eta \in \mathfrak{A}_0(\xi) ,$$

so that ξ is π-bounded and $\|\pi(\xi)\| \leq \gamma(\omega)\|\pi'(\xi')\|$ by Lemma 6.3. Making use of Lemma 6.3', we get similarly the assertion for $R(\omega)^T$. This completes the proof.

Now, let

$$\Delta = \int_0^\infty \lambda dE(\lambda) \tag{8.3}$$

be the spectral decomposition of Δ. For each bounded continuous function f on $[0,\infty]$, $f(\Delta)$ is given by:

$$f(\Delta) = \int_0^\infty f(\lambda)dE(\lambda) .$$

Take a function f in $A[0,\infty]$. Suppose the boundary Γ_f of U_f is a simply closed smooth curve. Then f is represented by the integral:

$$f(z) = \frac{i}{2\pi} \oint_{\Gamma_f} (z - \omega)^{-1} f(\omega) d\omega .$$

Therefore, $f(\Delta)$ is given by the integral:

$$f(\Delta) = \frac{i}{2\pi} \oint_{\Gamma_f} f(\omega)(\Delta - \omega)^{-1} d\omega . \tag{8.4}$$

Then the following is an immediate consequence of Lemma 8.1.

LEMMA 8.2. If f is in $A[0,\infty]$, then

$$f(\Delta)\mathfrak{A}' \subset \mathfrak{A}'' \text{ and } f(\Delta^{-1})\mathfrak{A} \subset \mathfrak{A}' \ ;$$

$$(8.5) \quad \|\pi(f(\Delta)\xi')\| \leq \frac{1}{2\pi}\gamma_f \sup_{\omega\in\Gamma_f} \gamma(\omega)|f(\omega)|\,\|\pi'(\xi')\|, \quad \xi' \in \mathfrak{A}' \ ;$$

$$(8.5') \quad \|\pi'(f(\Delta^{-1})\xi)\| \leq \frac{1}{2\pi}\gamma_f \sup_{\omega\in\Gamma_f} \gamma(\omega)|f(\omega)|\,\|\pi(\xi)\|, \quad \xi \in \mathfrak{A} \ ,$$

where γ_f is the length of Γ_f.

Define the sets $\mathfrak{A}^{\#}$ and $\mathfrak{A}^{\flat}$ by:

$$\mathfrak{A}^{\#} = \{\xi \in \mathfrak{A}' \cap \mathfrak{D}(\Delta);\ \Delta\xi \in \mathfrak{A}'\} \ ,$$

$$\mathfrak{A}^{\flat} = \{\xi \in \mathfrak{A} \cap \mathfrak{D}(\Delta^{-1});\ \Delta^{-1}\xi \in \mathfrak{A}\} \ .$$

Remark that ξ is in $\mathfrak{A}^{\#}$ if and only if ξ is in $\mathfrak{D}^{\#} \cap \mathfrak{A}'$ and $\xi^{\#}$ is in $\mathfrak{A}'$.

LEMMA 8.3. $\mathfrak{A}^{\#}$ is a self-adjoint subalgebra of $\mathfrak{A}$ and dense in the Hilbert space $\mathfrak{D}^{\#}$; similarly $\mathfrak{A}^{\flat}$ is a self-adjoint subalgebra of $\mathfrak{A}'$ and dense in the Hilbert space $\mathfrak{D}^{\flat}$. Furthermore, if ξ and η are in $\mathfrak{A}^{\#}$, then we have

$$(8.6) \qquad \Delta(\xi\eta) = (\Delta\xi)(\Delta\eta) \ ;$$

if ξ and η are in $\mathfrak{A}^{\flat}$, then

$$(8.6') \qquad \Delta^{-1}(\ \eta) = (\Delta^{-1}\ )(\Delta^{-1}\eta) \ .$$

<u>Proof</u>. If ξ is in $\mathfrak{A}^{\#}$, then $\eta = (1 + \Delta)\xi$ is in $\mathfrak{A}'$, so that $\xi = (1 + \Delta)^{-1}\eta$ is in $\mathfrak{A}$ by Lemma 8.1. Since

$$\eta^{b} = (\xi + \Delta\xi)^{b} = \xi^{b} + \xi^{\#} = (1 + \Delta)\xi^{\#} ,$$

$\xi^{\#} = (\eta - \xi)^{b}$ is in $\mathfrak{A}'$, and $\Delta\xi^{\#} = \eta^{b} - \xi^{\#}$ is in $\mathfrak{A}'$, so that $\xi^{\#}$ is also in $\mathfrak{A}^{\#}$. Hence $\mathfrak{A}^{\#}$ is self-adjoint under the involution $\xi \mapsto \xi^{\#}$. If ξ and η are both in $\mathfrak{A}^{\#}$, then we have, by the remark preceeding our lemma,

$$(\Delta\xi)(\Delta\eta) = \xi^{\#b}\eta^{\#b} = (\eta^{\#}\xi^{\#})^{b} = (\xi\eta)^{\#b} = \Delta(\xi\eta) .$$

Hence $\xi\eta$ and $\Delta(\xi\eta)$ are both in $\mathfrak{A}'$, so that $\xi\eta$ belongs to $\mathfrak{A}^{\#}$. Hence $\mathfrak{A}^{\#}$ is a subalgebra of $\mathfrak{A}$.

Now, we shall show the density of $\mathfrak{A}^{\#}$ in $\mathfrak{D}^{\#}$. By Lemma 1.1, it is sufficient to show that $(1 + \Delta^{\frac{1}{2}})\mathfrak{A}^{\#}$ is dense in $\mathfrak{H}$. Since the functions $z/(1 + z^2)$ and $1/(1 + z^2)$ are in $A[0,\infty]$, $\Delta^{-1}(1 + \Delta^{-2})^{-1}\mathfrak{A} \subset \mathfrak{A}'$ and $(1 + \Delta^{-2})^{-1}\mathfrak{A} \subset \mathfrak{A}'$, so that $\mathfrak{A}^{\#} \supset \Delta^{-1}(1 + \Delta^{-2})^{-1}\mathfrak{A}$; hence

$$(1 + \Delta^{\frac{1}{2}})\mathfrak{A}^{\#} \supset (1 + \Delta^{\frac{1}{2}})\Delta^{-1}(1 + \Delta^{-2})^{-1}\mathfrak{A} .$$

$\Delta^{-1}(1 + \Delta^{-2})^{-1}(1 + \Delta^{\frac{1}{2}}) = \Delta(1 + \Delta^{\frac{1}{2}})(1 + \Delta^{2})^{-1}$ is a bounded operator with dense range in $\mathfrak{H}$ and $\mathfrak{A}$ is dense in $\mathfrak{H}$, so that $(1 + \Delta^{\frac{1}{2}})\mathfrak{A}^{\#}$ is dense in $\mathfrak{H}$. The assertion for $\mathfrak{A}^{b}$ is proved by the symmetric arguments.

LEMMA 8.4. If the complex numbers ω_1, ω_2, and $\omega_1\omega_2$ are all in $\tilde{C} - [0,\infty]$ and if one of $\xi_1, \xi_2 \in \mathfrak{H}$ belongs to $\mathfrak{A}^{\#}$, then we have

$$(8.7)\quad \begin{cases} (R(\omega_1)\xi_1)(R(\omega_2)\xi_2) = R(\omega_1\omega_2)\{\xi_1\xi_2 + \omega_1(R(\omega_1)\xi_1)\xi_2 + \omega_2\xi_1(R(\omega_2)\xi_2)\} \\ \qquad = R(\omega_1\omega_2)\{(\Delta R(\omega_1)\xi_1)\xi_2 + \omega_2\xi_1(R(\omega_2)\xi_2)\} \\ \qquad = R(\omega_1\omega_2)\{\omega_1(R(\omega_1)\xi_1)\xi_2 + \xi_1(\Delta R(\omega_2)\xi_2)\} . \end{cases}$$

Proof. From the calculation:

$$(8.8)\quad \begin{cases} \Delta R(\omega)\xi = \Delta(\Delta - \omega)^{-1}\xi = \omega^{-1}(\omega^{-1} - \Delta^{-1})^{-1}\xi \\ R(\omega)\xi = \omega^{-1}(\Delta R(\omega)\xi - \xi),\ \xi \in \mathfrak{H} \text{ and } \omega \in \tilde{C} - [0,\infty] , \end{cases}$$

it follows by Lemma 8.1 that $R(\omega)\mathfrak{A}^{\#} \subset \mathfrak{A}^{\#}$ for each $\omega \in \tilde{C} - [0,\infty]$. Suppose ξ_1 belongs to $\mathfrak{A}^{\#}$. Then both sides of equality (8.7) are continuous functions of ξ_2, so we may assume that both ξ_1 and ξ_2 belong to $\mathfrak{A}^{\#}$, since $\mathfrak{A}^{\#}$ is dense in $\mathfrak{H}$. Then we have

$$\begin{aligned} \Delta((R(\omega_1)\xi_1)(R(\omega_2)\xi_2)) &= (\Delta R(\omega_1)\xi_1)(\Delta R(\omega_2)\xi_2) \quad \text{by (8.6)} \\ &= (\xi_1 + \omega_1 R(\omega_1)\xi_1)(\xi_2 + \omega_2 R(\omega_2)\xi_2) ; \end{aligned}$$

hence we have

$$\begin{aligned} (\Delta - \omega_1\omega_2)&\{(R(\omega_1)\xi_1)(R(\omega_2)\xi_2)\} \\ &= \xi_1\xi_2 + \omega_1(R(\omega_1)\xi_1)\xi_2 + \omega_2\xi_1(R(\omega_2)\xi_2) \\ &= (\Delta R(\omega_1)\xi_1)\xi_2 + \omega_2\xi_1(R(\omega_2)\xi_2) \\ &= \omega_1(R(\omega_1)\xi_1)\xi_2 + \xi_1(\Delta R(\omega_2)\xi_2) , \end{aligned}$$

which completes the proof.

Define the projection E_γ for $\gamma > 1$ by:

(8.9) $$E_\gamma = \int_{1/\gamma}^{\gamma} dE(\lambda) .$$

Then it is clear that

(8.10) $$E_\gamma^T = E_\gamma \; ; \quad \|E_\gamma \Delta\| \leq \gamma \; ; \quad \|E_\gamma \Delta^{-1}\| \leq \gamma .$$

Let f be a function analytic in a neighborhood U_f of $[0,\infty]$. We assume that the boundary Γ_f of U_f is a simply closed smooth curve. Consider the set

(8.11) $$W_f = \{tz^{-1} : 0 \leq t \leq \infty, \; z \in U_f^C \cup \Gamma_f\} .$$

Then W_f and $[\frac{1}{\gamma},\gamma]$, $\gamma > 1$, are disjoint compact sets in $\tilde{C}$. The following picture will illustrate the location of W_f and $[\frac{1}{\gamma},\gamma]$:

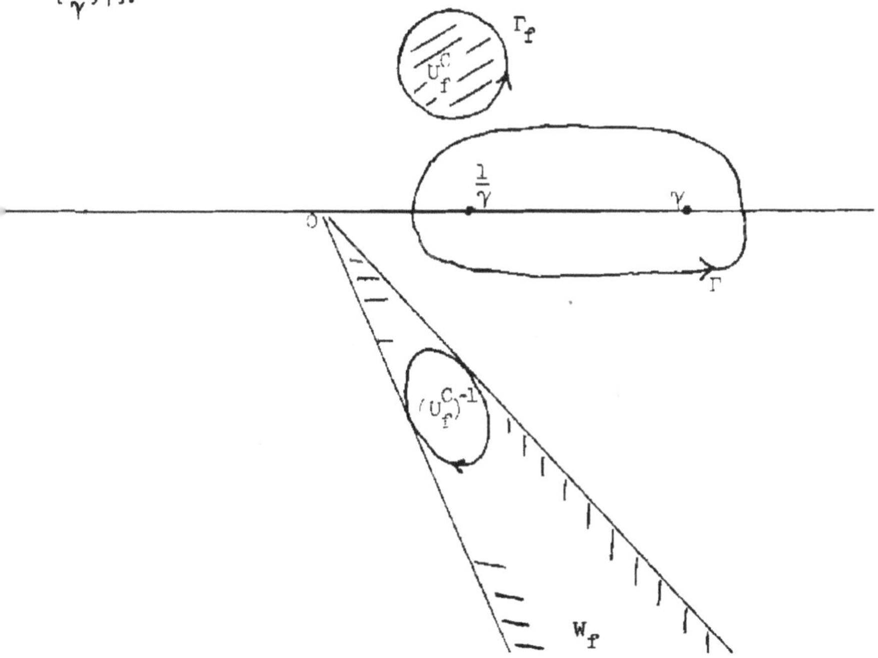

Take a simply closed smooth curve Γ whose interior contains $[\frac{1}{\gamma},\gamma]$, and whose exterior contains W_f and which meets the half line $[0,\infty]$ only at two points.

LEMMA 8.5. In the above situation, $f(\Delta)\mathfrak{A}^{\#} \subset \mathfrak{A}'$ and if ξ is in $\mathfrak{A}^{\#}$ then

$$(E_\gamma\eta)(f(\Delta)\xi) = \frac{i}{2\pi} \oint_\Gamma f(\omega^{-1}\Delta)[(R(\omega)E_\gamma\eta)\xi]d\omega \tag{8.12}$$

for every $\eta \in \mathfrak{H}$.

<u>Proof</u>. From the equality:

$$E_\gamma = \frac{i}{2\pi} \oint_\Gamma R(\omega)E_\gamma d\omega ,$$

equality (8.12) follows if f is a constant function. Considering the decomposition $f(z) = f(z) - f(\infty) + f(\infty)$, we may assume that f is in $A[0,\infty]$. Suppose ξ is in $\mathfrak{A}^{\#}$. By equality (8.8), $R(\omega)\xi$ belongs to $\mathfrak{A}^{\#}$. Furthermore, the inequality:

$$\|\pi'(R(\omega)\xi)\| = \|\omega^{-1}\pi'(\omega^{-1}R(\omega^{-1})^T\xi - \xi)\|$$

$$\leq \frac{1}{|\omega|^2}\gamma(\omega^{-1})\|\pi(\xi)\| + \frac{1}{|\omega|}\|\pi'(\xi)\|$$

implies that $\|\pi'(R(\omega)\xi)\|$ is uniformly bounded along the curve Γ_f, so that $f(\Delta)\xi$ is π'-bounded because the range of $(f(\Delta) - f(0)\cdot 1)$ is contained in $\mathfrak{D}(\Delta^{-1})$ whenever f is in $A[0,\infty]$. Therefore, the first assertion follows.

In Lemma 8.4, fix ω_2 on the curve Γ_f and replace ξ_1 and

ξ_2 by $E_\gamma\eta$ and ξ respectively. Then (8.7) turns out to be:

$$(R(\omega_1)E_\gamma\eta)(R(\omega_2)\xi) = R(\omega_1\omega_2)\{\omega_1(R(\omega_1)E_\gamma\eta)\xi + (E_\gamma\eta)(\Delta R(\omega_2)\xi)\}$$

Multiplying by $(\frac{i}{2\pi})$ on both sides and integrating with respect to ω_1 along the curve Γ, we get the left hand side:

$$(E_\gamma\eta)(R(\omega_2)\xi) = \frac{i}{2\pi}\oint_\Gamma (R(\omega_1)E_\gamma\eta)(R(\omega_2)\xi)d\omega_1 .$$

Noticing that $\omega_1\omega_2 \notin [0,\infty]$ for every $\omega_1 \in \Gamma$ and $\omega_2 \in \Gamma_f$ we conclude that $R(\omega_1\omega_2)$ is an analytic function of ω_1 on and inside of the curve Γ, so that the second term of the right hand side turns out to be:

$$\frac{i}{2\pi}\int_\Gamma R(\omega_1\omega_2)(E_\gamma\eta)(\Delta R(\omega_2)\xi)d\omega_1 = 0 .$$

Therefore, we have

$$(8.13)\quad (E_\gamma\eta)(R(\omega_2)\xi) = \frac{i}{2\pi}\oint_\Gamma \omega_1 R(\omega_1\omega_2)[(R(\omega_1)E_\gamma\eta)\xi]d\omega_1 .$$

Multiplying by $f(\omega_2)$ on both sides of (8.13) and integrating with respect to ω_2 along the curve Γ_f, we get

$$(E_\gamma\eta)(f(\Delta)\xi) = \frac{-1}{(2\pi^2)}\oint_{\Gamma_f}\oint_\Gamma f(\omega_2)\omega_1 R(\omega_1\omega_2)[(R(\omega_1)E_\gamma\eta)\xi]d\omega_1 d\omega_2 .$$

Noticing that

$$f(\omega_1^{-1}t) = \frac{i}{2\pi} \oint_{\Gamma_f} f(\omega_2)\omega_1(t - \omega_1\omega_2)^{-1}d\omega_2$$

for $\omega_1 \in \Gamma$ and $t \in [0,\infty]$, we get

$$f(\omega_1^{-1}\Delta) = \frac{i}{2\pi} \oint_{\Gamma_f} f(\omega_2)\omega_1 R(\omega_1\omega_2)d\omega_2 \ ;$$

and then we get

$$(E_\gamma\eta)(f(\Delta)\xi) = \frac{i}{2\pi} \oint_{\Gamma} f(\omega^{-1}\Delta)[(R(\omega)E_\gamma\eta)\xi]d\omega \ ,$$

which completes the proof.

9. The One-Parameter Automorphisms defined by the Modular Operator Δ

Now, we are at the most crucial point in the proof of the fundamental theorem of Tomita. Consider the negative half line $W = \{z \in \tilde{C} : -z \in [0,\infty]\}$. For each complex number α, define an analytic function z^α on $\tilde{C} - W$ by

$$z^\alpha = \exp(\alpha \log|z| + i\alpha \arg z) \ ,$$

where $-\pi < \arg z < \pi$. Then we have $z^{\alpha+\beta} = z^\alpha \cdot z^\beta$. The normal operator Δ^α is given by:

$$(9.1) \qquad \Delta^\alpha = \int_0^\infty z^\alpha dE(z) = \exp(\alpha \log \Delta) \ .$$

If $\alpha = s + it$ for $s,t \in R$, we have

$$\Delta^{\alpha} = \Delta^{s}\Delta^{it} ;$$

Δ^s is a positive self-adjoint operator and Δ^{it} is unitary.

LEMMA 9.1. For each complex number α, $\mathfrak{A}^{\#} \cap \mathfrak{D}(\Delta^{\alpha})$ is dense in the Hilbert space $\mathfrak{D}(\Delta^{\alpha})$.

<u>Proof</u>. Let $\alpha = s + it$, $s,t \in R$. Take an integer $n \geq |s| + 1$ and consider the function $f(z) = z^n(z^{2n} + 1)^{-1}$. We claim that $\Delta^{-1}f(\Delta^{-1})\mathfrak{A}^{\flat} \subset \mathfrak{A}^{\#} \cap \mathfrak{D}(\Delta^{\alpha})$. Indeed, if η belongs to $\mathfrak{A}^{\flat}$, then η and $\Delta^{-1}\eta$ both belong to $\mathfrak{A}$, so that $f(\Delta^{-1})\eta$ and $\Delta^{-1}f(\Delta^{-1})\eta$ both belong to $\mathfrak{A}'$, because $f(z)$ and $zf(z)$ are both in $A[0,\infty]$; hence $\Delta^{-1}f(\Delta^{-1})\eta \in \mathfrak{A}^{\#}$. By definition of the function f, the range of $\Delta^{-1}f(\Delta^{-1})$ is contained in $\mathfrak{D}(\Delta^s) = \mathfrak{D}(\Delta^{\alpha})$.

Since $(1 + \Delta^s)\Delta^{-1}f(\Delta^{-1}) = \Delta^{n-1}(1 + \Delta^s)(1 + \Delta^{2n})^{-1}$ is a bounded self-adjoint operator with dense range in $\mathfrak{H}$, and $\mathfrak{A}^{\flat}$ is dense in $\mathfrak{H}$, $(1 + \Delta^s)\Delta^{-1}f(\Delta^{-1})\mathfrak{A}^{\flat}$ is dense in $\mathfrak{H}$; therefore by Lemma 1.1 $\Delta^{-1}f(\Delta^{-1})\mathfrak{A}^{\flat}$ is dense in the Hilbert space $\mathfrak{D}(\Delta^s) = \mathfrak{D}(\Delta^{\alpha})$, so that $\mathfrak{A}^{\#} \cap \mathfrak{D}(\Delta^{\alpha})$ is dense in $\mathfrak{D}(\Delta^{\alpha})$.

THEOREM 9.1. For each $\xi \in \mathfrak{A} \cap \mathfrak{D}(\Delta^{-\alpha})$ and $\eta \in \mathfrak{A}' \cap \mathfrak{D}(\Delta^{\alpha})$, $(\Delta^{-\alpha}\xi)\eta$ belongs to $\mathfrak{D}(\Delta^{\alpha})$ and satisfies the formula:

$$\Delta^{\alpha}(\Delta^{-\alpha}\xi)\eta = \xi(\Delta^{\alpha}\eta) . \tag{9.2}$$

<u>Proof</u>. Consider the function

$$f_{\delta}(z) = \left(\frac{z + \delta}{1 + \delta z}\right)^{\alpha}, \quad 0 \leq \delta \leq 1 . \tag{9.3}$$

Then $f_0(z) = z^{\alpha}$. If $0 < \delta \leq 1$, then f_δ is analytic except on the segment $-1/\delta \leq z \leq -\delta$. Hence f_δ is analytic in a neighborhood of $[0,\infty]$ and we may assume that the set W_{f_δ} defined by (8.11) is contained in the left half plane $\operatorname{Re} z \leq 0$. Now we shall apply Lemma 8.5 to f_δ. Take a simply closed smooth curve Γ as in the proof of Lemma 8.5, so we have the following picture:

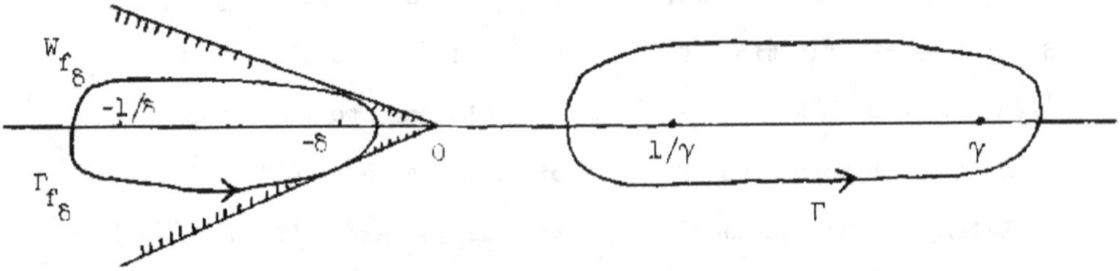

Take η_1 and η_2 in $\mathfrak{A}^{\#}$ and ξ in $\mathfrak{A}$. Since $E_\gamma\eta$ belongs to $\mathfrak{D}(\Delta^{\frac{1}{2}}) = \mathfrak{D}^{\#}$, $(E_\gamma\xi)^{\#}$ is defined. By Lemma 8.5, we get

$$\begin{aligned}(f_\delta(\Delta)\eta_1 | (E_\gamma\xi)^{\#}\eta_2) &= ((E_\gamma\xi)(f_\delta(\Delta)\eta_1) | \eta_2) \\ &= \frac{i}{2\pi}\oint_\Gamma (f_\delta(\omega^{-1}\Delta)[(R(\omega)E_\gamma\xi)\eta_1] | \eta_2)d\omega \\ &= \frac{i}{2\pi}\oint_\Gamma ((R(\omega)E_\gamma\xi)\eta_1 | f_\delta(\omega^{-1}\Delta)^*\eta_2)d\omega \\ &= \frac{i}{2\pi}\oint_\Gamma (\eta_1 | (R(\omega)E_\gamma\xi)^{\#}(f_\delta(\omega^{-1}\Delta)^*\eta_2))d\omega ,\end{aligned}$$

where the last step of the above calculation is justified as follows: the function $z \to (\omega\delta + z)^{\bar{\alpha}}(\bar{\omega} + \delta z)^{-\bar{\alpha}}$ is analytic in a neighborhood of $[0,\infty]$ and

$$f_\delta(\omega^{-1}\Delta)^* = (\bar{\omega}\delta + \Delta)^{\bar{\alpha}}(\bar{\omega} + \delta\Delta)^{-\bar{\alpha}} ;$$

hence Lemma 8.5 assures that $f_\delta(\omega^{-1}\Delta)^*\eta_2$ is in $\mathfrak{A}'$, and also $R(\omega)E_\gamma\xi$ belongs to $\mathfrak{D}^{\#}$. Therefore, we get

$$(9.4)\quad (f_\delta(\Delta)\eta_1|(E_\gamma\xi)^{\#}\eta_2) = \frac{i}{2\pi}\oint_\Gamma (\eta_1|(R(\omega)E_\gamma\xi)^{\#}(f_\delta(\omega^{-1}\Delta)^*\eta_2))d\omega$$

for each $\xi \in \mathfrak{A}$ and $\eta_1,\eta_2 \in \mathfrak{A}^{\#}$.

Since

$$f_\delta(\omega^{-1}\Delta)^* = (1+\bar{\omega}\delta\Delta^{-1})^{\bar{\alpha}}(\delta+\bar{\omega}\Delta^{-1})^{-\bar{\alpha}},$$

if we define a function g_δ^ω by:

$$g_\delta^\omega(z) = \left(\frac{1+\bar{\omega}\delta z}{\delta+\bar{\omega}z}\right)^{\bar{\alpha}} - \delta^{\bar{\alpha}},$$

then

$$f_\delta(\omega^{-1}\Delta)^* = g_\delta^\omega(\Delta^{-1}) + \delta^{\bar{\alpha}}\cdot 1$$

and g_δ^ω belongs to $A[0,\infty]$. Hence by (8.5'), we have

$$\|\pi'(f_\delta(\omega^{-1}\Delta)^*\eta_2)\| \le |\delta^{\bar{\alpha}}|\,\|\pi'(\eta_2)\| + \frac{1}{2\pi}\,\ell \sup_{z\in\Gamma_{f_\delta}} (\gamma(z)|g_\delta^\omega(z)|)\,\|\pi(\eta_2)\|,$$

where Γ_{f_δ} denotes a simply closed smooth curve with length ℓ, whose interior contains the segment $[-\frac{1}{\delta},-\delta]$.

Hence there exists a constant γ_0 not depending on ω such that

$$\|\pi'(f_\delta(\omega^{-1}\Delta)^*\eta_2)\| \le \gamma_0 .$$

The operator: $\xi \to (E_\gamma\xi)^{\#}$ is bounded, so that the function: $\omega \to (R(\omega)E_\gamma\xi)^{\#}$ is continuous on Γ; hence the function:

$$\omega \in \Gamma \mapsto (R(\omega)E_\gamma\xi)^{\#}(f_\delta(\omega^{-1}\Delta)^*\eta_2) \in \mathfrak{H}$$

is bounded on Γ. Thus both sides of equality (9.4) are continuous functions of η_1, so that equality (9.4) remains true for every $\eta_1 \in \mathfrak{H}$. Thus we get

$$(9.5)\quad (f_\delta(\Delta)\eta_1|(E_\gamma\xi)^{\#}\eta_2) = \frac{i}{2\pi}\oint_\Gamma ((R(\omega)E_\gamma\xi)\eta_1|f_\delta(\omega^{-1}\Delta)^*\eta_2)d\omega$$

for $\eta_1 \in \mathfrak{A}' \cap \mathfrak{D}(\Delta^\alpha)$, $\eta_2 \in \mathfrak{A}^{\#} \cap \mathfrak{D}(\Delta^\alpha)$ and $\xi \in \mathfrak{A} \cap \mathfrak{D}(\Delta^{-\alpha})$.

We shall show that equality (9.5) is also true in the case when $\delta = 0$. Write $\alpha = s + it$ and put

$$\gamma_1 = \inf\{\mathrm{Re}\ \omega;\ \bar{\omega} \in \Gamma\},$$

$$\gamma_2 = \sup\{|\omega|;\ \omega \in \Gamma\},$$

$$\gamma_3 = \sup\{|\arg\omega|;\ \bar{\omega} \in \Gamma\},$$

$$\gamma_4 = \exp(\gamma_3|t|).$$

For each $\omega \in \Gamma$ and $\lambda \in [0,\infty]$, we have

$$\begin{aligned} |f_\delta(\omega^{-1}\lambda)^\alpha| &= \left|\left(\frac{\lambda+\omega\delta}{\omega+\delta\lambda}\right)^\alpha\right| \\ &= \left|\frac{\lambda+\omega\delta}{\omega+\delta\lambda}\right|^s \exp\left\{-t\arg\left(\frac{\lambda+\omega\delta}{\omega+\delta\lambda}\right)\right\} \\ &= \left|\frac{\lambda+\omega\delta}{\omega+\delta\lambda}\right|^s \exp\{t(\arg(\omega+\delta\lambda) - \arg(\lambda+\omega\delta))\}. \end{aligned}$$

Since $\mathrm{Re}\ \omega > 0$ and λ is real, we have

$$|\arg(\omega+\delta\lambda) - \arg(\lambda+\omega\delta)| \le |\arg\omega|,$$

and then

$$\exp\{t(\arg(\omega+\delta\lambda) - \arg(\lambda+\omega\delta))\} \le \exp(\gamma_3 t) \le \gamma_4.$$

If $s > 0$, then since $|1+\delta\omega^{-1}\lambda| \ge 1$, we have

$$\left|\frac{\lambda + \omega\delta}{\omega + \delta\lambda}\right|^s \leq |\omega^{-1}\lambda + \delta|^s \leq (|\omega|^{-1}\lambda + \delta)^s$$

$$\leq (\gamma_1^{-1}\lambda + \delta)^s \leq (\gamma_1^{-1}\lambda + 1)^s .$$

Noticing that $\mathfrak{D}(\Delta^s) = \mathfrak{D}((\gamma_1^{-1}\Delta + 1)^s)$, we have, for each $\eta \in \mathfrak{D}(\Delta^s)$,

$$\|f_\delta(\omega^{-1}\Delta)^*\eta\| \leq \gamma_4\|(\gamma_1^{-1}\Delta + 1)^s\eta\| .$$

If $s < 0$, then we have

$$\left|\frac{\lambda + \omega\delta}{\omega + \delta\lambda}\right|^s = \left|\frac{1 + \lambda^{-1}\omega\delta}{\lambda^{-1}\omega + \delta}\right|^s \leq |\lambda^{-1}\omega + \delta|^{-s}$$

$$\leq (\gamma_2\lambda^{-1} + \delta)^{-s} \leq (\gamma_2\lambda^{-1} + 1)^{-s} .$$

Noticing again that $\mathfrak{D}(\Delta^s) = \mathfrak{D}((\gamma_2\Delta^{-1} + 1)^{-s})$, we have

$$\|f_\delta(\omega^{-1}\Delta)^*\eta\| \leq \gamma_4\|(\gamma_2\Delta^{-1} + 1)^{-s}\eta\|, \quad \eta \in \mathfrak{D}(\Delta^s) .$$

If $s = 0$, then

$$\|f_\delta(\omega^{-1}\Delta)^*\eta\| \leq \gamma_4\|\eta\| .$$

Since η_2 belongs to $\mathfrak{D}(\Delta^\alpha)$ $(= \mathfrak{D}(\Delta^s))$, for every $\varepsilon > 0$ we can find a $\lambda_0 > 1$ such that

$$\gamma_4\|(\gamma_1^{-1}\Delta + 1)^s(1 - E_{\lambda_0})\eta_2\| < \varepsilon\|\eta_2\|$$

in the case when $s > 0$, or

$$\gamma_4\|(\gamma_2\Delta^{-1} + 1)^{-s}(1 - E_{\lambda_0})\eta_2\| < \varepsilon\|\eta_2\|$$

in the case when $s \leq 0$. Hence we have

$$\|f_\delta(\omega^{-1}\Delta)^*(1 - E_{\lambda_0})\eta_2\| < \varepsilon\|\eta_2\| .$$

Remark that the number $\lambda_0 > 0$, hence the projection E_{λ_0}, depends on neither $\omega \in \Gamma$ nor $\delta < 1$. Since $\bar{f}_\delta(\bar{\omega}^{-1}\lambda)$ converges to $\bar{f}_0(\bar{\omega}^{-1}\lambda)$ uniformly for $\lambda \in [1/\lambda_0, \lambda_0]$ and $\omega \in \Gamma$ as δ tends to zero, we have

$$\sup_{\omega\in\Gamma} \|f_\delta(\omega^{-1}\Delta)^* E_{\lambda_0}\eta_2 - \bar{\omega}^{-\bar{\alpha}}\Delta^{\bar{\alpha}} E_{\lambda_0}\eta_2\| \leq \varepsilon\|\eta_2\|$$

for a sufficiently small $\delta > 0$. Thus we conclude that

$$\lim_{\delta\to 0}\sup_{\omega\in\Gamma} \|f_\delta(\omega^{-1}\Delta)^*\eta_2 - \bar{\omega}^{-\bar{\alpha}}\Delta^{\bar{\alpha}}\eta_2\| = 0$$

for each $\eta_2 \in \mathfrak{D}(\Delta^{\bar{\alpha}})$. Of course, it is an immediate consequence of the above discussion that

$$\lim_{\delta\to 0} \|f_\delta(\Delta)\eta_1 - \Delta^\alpha\eta_1\| = 0$$

for each $\eta_1 \in \mathfrak{D}(\Delta^\alpha)$. Therefore, equality (9.5) turns out, by taking limits, to be:

$$(9.6)\quad (\Delta^\alpha\eta_1|(E_\gamma\xi)^\#\eta_2) = \frac{i}{2\pi}\oint_\Gamma ((R(\omega)E_\gamma\xi)\eta_1|\bar{\omega}^{-\bar{\alpha}}\Delta^{\bar{\alpha}}\eta_2)d\omega$$

for each $\eta_1 \in \mathfrak{A}' \cap \mathfrak{D}(\Delta^\alpha)$, $\eta_2 \in \mathfrak{A}^\# \cap \mathfrak{D}(\Delta^{\bar{\alpha}})$ and $\xi \in \mathfrak{A} \cap \mathfrak{D}(\Delta^{-\alpha})$. The right hand side of (9.6) equals:

$$\frac{i}{2\pi}\oint_\Gamma (\omega^{-\alpha}(R(\omega)E_\gamma\xi)\eta_1|\Delta^{\bar{\alpha}}\eta_2)d\omega = ((\Delta^{-\alpha}E_\gamma\xi)\eta_1|\Delta^{\bar{\alpha}}\eta_2) .$$

Hence we get

$$(\Delta^{\alpha}\eta_1|(E_\gamma\xi)^{\#}\eta_2) = ((\Delta^{-\alpha}E_\gamma\xi)\eta_1|\Delta^{\bar{\alpha}}\eta_2) .$$

Since we have by (8.9)

$$(E_\gamma\xi)^{\#} = SE_\gamma\xi = \Delta^{-\frac{1}{2}}JE_\gamma\xi = \Delta^{-\frac{1}{2}}E_\gamma^T J\xi$$

$$= E_\gamma\Delta^{-\frac{1}{2}}J\xi = E_\gamma\xi^{\#} ,$$

we have

$$\lim_{\gamma\to\infty} (E_\gamma\xi)^{\#} = \xi^{\#} .$$

Recalling that η_2 is in $\mathfrak{A}'$, we get

$$\lim_{\gamma\to\infty} (E_\gamma\xi)^{\#}\eta_2 = \xi^{\#}\eta_2 .$$

Since ξ belongs to $\mathfrak{D}(\Delta^{-\alpha})$, we have

$$\lim_{\gamma\to\infty} (\Delta^{-\alpha}E_\gamma\xi) = \Delta^{-\alpha}\xi ;$$

$$\lim_{\gamma\to\infty} (\Delta^{-\alpha}E_\gamma\xi)\eta_1 = (\Delta^{-\alpha}\xi)\eta_1$$

because η_1 belongs to $\mathfrak{A}'$. Therefore, we have

$$((\Delta^{-\alpha}\xi)\eta_1|\Delta^{\bar{\alpha}}\eta_2) = (\Delta^{\alpha}\eta_1|\xi^{\#}\eta_2)$$

(9.7)

$$= (\xi(\Delta^{\alpha}\eta_1)|\eta_2).$$

By Lemma 9.1, $\mathfrak{A}^{\#} \cap \mathfrak{D}(\Delta^{\alpha})$ is dense in the Hilbert space $\mathfrak{D}(\Delta^{\alpha}) = \mathfrak{D}(\Delta^{\bar{\alpha}})$, so that equality (9.7) implies that $(\Delta^{-\alpha}\xi)\eta_1$ belongs to $\mathfrak{D}(\Delta^{\alpha})$ and that

$$\Delta^{\alpha}((\Delta^{-\alpha}\xi)\eta_1) = \xi(\Delta^{\alpha}\eta_1)$$

for each $\xi \in \mathfrak{A} \cap \mathfrak{D}(\Delta^{-\alpha})$ and $\eta_1 \in \mathfrak{A}' \cap \mathfrak{D}(\Delta^{\alpha})$. This completes the proof.

COROLLARY 9.1. Δ^{it}, $-\infty < t < \infty$, forma a one-parameter automorphism group of $\mathfrak{A}$ and

$$\pi(\Delta^{it}\xi) = \Delta^{it}\pi(\xi)\Delta^{-it}, \quad \xi \in \mathfrak{A} . \tag{9.8}$$

Proof. Noticing that $\mathfrak{D}(\Delta^{it}) = \mathfrak{D}(\Delta^{-it}) = \mathfrak{H}$, we get

$$\Delta^{it}(\xi(\Delta^{-it}\eta)) = (\Delta^{it}\xi)\eta, \quad \xi \in \mathfrak{A}, \quad \eta \in \mathfrak{A}' .$$

Hence we have $\pi(\Delta^{it}\xi) = \Delta^{it}\pi(\xi)\Delta^{-it}$. Of course, $\Delta^{it}\xi$ is π-bounded. This completes the proof.

Therefore, the one-parameter unitary group $\{\Delta^{it}\}$ induces a one-parameter *-automorphism group of the left von Neumann algebra $\mathfrak{L}(\mathfrak{A})$ of $\mathfrak{A}$.

10. Formulation of the Modular Hilbert Algebra

Now, we are at the final stage of the construction of the modular Hilbert algebra equivalent to a given generalized Hilbert algebra. Throughout this section, we suppose $\mathfrak{A}$ is an achieved generalized Hilbert algebra.

LEMMA 10.1. If f is a continuous positive definite function of a real variable, then $f(\log \Delta)$ carries $\mathfrak{A}$ into $\mathfrak{A}$ and

$$(f(\log \Delta)\xi)^{\#} = f(\log \Delta)\xi^{\#}, \quad \xi \in \mathfrak{A} . \tag{10.1}$$

Proof By Bochner's Theorem, f is represented by an integral:

$$f(\lambda) = \int_{-\infty}^{\infty} e^{i\lambda t}\, d\mu(t) ,$$

where μ is a finite positive measure. Hence we have

$$f(\log \Delta) = \int_{-\infty}^{\infty} \Delta^{it} d\mu(t) ,$$

where the integral is taken in the strong operator topology. Therefore, for each $\xi \in \mathfrak{A}$ and $\eta \in \mathfrak{A}'$, we have

$$\begin{aligned} \|(f(\log \Delta)\xi)\eta\| &= \left\| \int_{-\infty}^{\infty} (\Delta^{it}\xi)\eta \; d\mu(t) \right\| \\ &\leqq \int_{-\infty}^{\infty} \|(\Delta^{it}\xi)\eta\| d\mu(t) \\ &\leqq \int_{-\infty}^{\infty} \|\pi(\xi)\| \|\eta\| d\mu(t) = f(0) \|\pi(\xi)\| \|\eta\| , \end{aligned}$$

so that $f(\log \Delta)\xi$ belongs to $\mathfrak{A}$, because $f(\log \Delta)$ leaves the domain $\mathfrak{D}(\Delta^{\frac{1}{2}})$ invariant. For each $\xi \in \mathfrak{A}$, we have

$$\begin{aligned} (f(\log \Delta)\xi)^{\#} &= Sf(\log \Delta)\xi = \Delta^{-\frac{1}{2}} Jf(\log \Delta)\xi \\ &= \Delta^{-\frac{1}{2}} f(\log \Delta^{-1}) J\xi = \Delta^{-\frac{1}{2}} f(-\log \Delta) J\xi \\ &= \Delta^{-\frac{1}{2}} f(\log \Delta) J\xi = f(\log \Delta) \Delta^{-\frac{1}{2}} J\xi \\ &= f(\log \Delta)\xi^{\#} . \end{aligned}$$

Now, Tomita's fundamental theorem is at hand as follows:

THEOREM 10.1. For every generalized Hilbert algebra $\mathfrak{A}$, there exists a modular Hilbert algebra $\mathfrak{B}$ which is equivalent to $\mathfrak{A}$. Therefore, there is an isometric involution J of the Hilbert space $\mathfrak{H}$, which is the completion of $\mathfrak{A}$, such that

$$(10.2) \qquad J\mathfrak{L}(\mathfrak{A})J = \mathfrak{L}(\mathfrak{A})' .$$

In particular, the left von Neumann algebra $\mathfrak{L}(\mathfrak{A})$ of $\mathfrak{A}$ is anti-isomorphic to its commutant $\mathfrak{L}(\mathfrak{A})'$.

Proof. We may assume that $\mathfrak{A}$ is achieved. Keep the notation as before. Let $\mathcal{E}$ denote the linear space spanned by $f * g$, where f and g are continuous functions of real variables with compact support and

$$f * g(t) = \int_{-\infty}^{\infty} f(s)g(t - s)ds .$$

For each function f of a real variable, put $\tilde{f}(t) = \overline{f(-t)}$. Putting $e_\alpha(t) = e^{\alpha t}$ for each complex number α, we have

$$e_\alpha(f * g) = (e_\alpha f) * (e_\alpha g) ,$$

so that the map: $f \to e_\alpha f$ carries $\mathcal{E}$ onto $\mathcal{E}$. Let $\mathfrak{B}$ be the subalgebra of $\mathfrak{A}$ generated algebraically by $f(\log \Delta)\xi$, $f \in \mathcal{E}$ and $\xi \in \mathfrak{A}$. Since

$$(f(\log \Delta)\xi)^\# = \tilde{f}(\log \Delta)\xi^\# ,$$

$\mathfrak{B}$ is a self-adjoint subalgebra of $\mathfrak{A}$. By the equality:

$$\Delta^{\alpha} f(\log \Delta)\xi = e_{\alpha}(\log \Delta) f(\log \Delta)\xi$$

$$= (e_{\alpha} f)(\log \Delta)\xi ,$$

$\mathfrak{B}$ is invariant under the operator Δ^{α}.

To complete the proof, we shall show that $\mathfrak{B}$ is dense in the Hilbert space $\mathfrak{D}(\Delta^{\alpha})$. Put $\alpha = s + it$. Take an arbitrary ξ in $\mathfrak{D}(\Delta^{\alpha})$ and $\varepsilon > 0$. Then there exists a continuous function f with compact support such that

$$\|(1 + \Delta^{s})(f(\log \Delta)\xi - \xi)\| < \varepsilon .$$

We can find a sequence $\{f_n\}$ in $\mathcal{E}$ such that $\{f_n\}$ converges to f uniformly and supp. f_n is contained in a fixed compact set K in R. Let χ_K be the characteristic function of K. Then $(1 + \Delta^{s})\chi_K(\log \Delta)$ is bounded and

$$(1 + \Delta^{s}) f_n(\log \Delta) = (1 + \Delta^{s})\chi_K(\log \Delta) f_n(\log \Delta)\xi .$$

Hence we have

$$\lim_{n\to\infty} \|(1 + \Delta^{s})(f_n(\log \Delta)\xi - f(\log \Delta)\xi)\| = 0 .$$

Hence we can find a function f_0 in $\mathcal{E}$ such that

$$\|(1 + \Delta^{s})(f_0(\log \Delta)\xi - \xi)\| < \varepsilon ,$$

which means by Lemma 1.1 that $\mathfrak{B}$ is dense in $\mathfrak{D}(\Delta^{\alpha})$. In particular, $\mathfrak{B}$ is dense in $\mathfrak{D}(\Delta^{\frac{1}{2}})$, so that $\mathfrak{B}'' = \mathfrak{A}$ and $\mathfrak{B}' = \mathfrak{A}'$ by Lemma 5.2. Therefore, $\mathfrak{B}$ satisfies all postulates for a modular Hilbert algebra, except possibly postulate (VII), and is equivalent to $\mathfrak{A}$.

Now we shall show that the function: $\alpha \to (\Delta^{\alpha}\xi|\eta)$, $\xi,\eta \in \mathfrak{B}$, is analytic. Consider the set $\mathfrak{C}$ of all elements ξ with the following properties:

(a) ξ is in $\bigcap_\alpha \mathfrak{D}(\Delta^\alpha)$;

(b) $\Delta^\alpha \xi$ is in $\mathfrak{A}^\#$ for every complex number α;

(c) the function: $\alpha \to (\Delta^\alpha \xi | \eta)$ is analytic for each $\eta \in \mathfrak{A}$.

Then $\mathfrak{C}$ clearly contains all elements of the form $f(\log \Delta)\xi$, $f \in \mathcal{E}$ and $\xi \in \mathfrak{A}$. We shall show that $\mathfrak{C}$ is an algebra. Take ξ_1 and ξ_2 in $\mathfrak{C}$ and $\eta \in \mathfrak{A}$. Then we have

$$\frac{1}{h} \{(\Delta^{\alpha+h}(\xi_1\xi_2)|\eta) - (\Delta^\alpha(\xi_1\xi_2)|\eta)\}$$

$$= \frac{1}{h} (\{(\Delta^{\alpha+h}\xi_1)(\Delta^{\alpha+h}\xi_2) - (\Delta^\alpha \xi_2)(\Delta^\alpha \xi_2)\}|\eta)$$

$$= \frac{1}{h} ((\Delta^{\alpha+h}\xi_1 - \Delta^\alpha\xi_1)\Delta^{\alpha+h}\xi_2|\eta) + \frac{1}{h} ((\Delta^\alpha\xi_1)(\Delta^{\alpha+h}\xi_2 - \Delta^\alpha\xi_2)|\eta)$$

$$= \frac{1}{h} (\Delta^{\alpha+h}\xi_1 - \Delta^\alpha\xi_1|\eta(\Delta^{\alpha+h}\xi_2)^\flat) + \frac{1}{h} (\Delta^{\alpha+h}\xi_2 - \Delta^\alpha\xi_2|(\Delta^\alpha\xi_1)^\#\eta)$$

$$= \frac{1}{h} (\Delta^{\alpha+h}\xi_1 - \Delta^\alpha\xi_1|\eta(J\Delta^{\alpha-\frac{1}{2}+h}\xi_2)) + \frac{1}{h} (\Delta^{\alpha+h}\xi_2 - \Delta^\alpha\xi_2|(\Delta^\alpha\xi_1)^\#\eta) .$$

As in the proof of Lemma 2.1, the functions: $\alpha \to \Delta^\alpha \xi_i$, $i = 1,2$, are strongly analytic, so that $\Delta^{\alpha-\frac{1}{2}+h}\xi_2$ converges to $\Delta^{\alpha-\frac{1}{2}}$ as h tends to 0; hence the function $\alpha \mapsto (\Delta^\alpha(\xi_1\xi_2)|\eta)$ is analytic. Hence $\xi_1\xi_2$ satisfies condition (c) as well as conditions (a) and (b), which means that $\mathfrak{C}$ is an algebra as desired. Therefore $\mathfrak{C}$ contains $\mathfrak{B}$. This completes the proof.

COROLLARY 10.1. If $\mathfrak{A}$ is a generalized Hilbert algebra, then we have

(10.3) $$J\mathfrak{A}'' = \mathfrak{A}' \quad \text{and} \quad J\mathfrak{A}' = \mathfrak{A}'' .$$

Proof. Let $\mathfrak{A}_0$ be a modular Hilbert algebra in $\mathfrak{A}''$ constructed in Theorem 10.1. Then we have, for each pair ξ, η in $\mathfrak{A}_0$,

$$J(\xi\eta) = (J\eta)(J\xi) .$$

Take an arbitrary $\eta \in \mathfrak{A}'$. Then we can choose a sequence $\{\eta_n\}$ in $\mathfrak{A}_0$ with $\eta = \lim_{n\to\infty} \eta_n$, so that we have for each $\xi \in \mathfrak{A}_0$,

$$\begin{aligned}(J\eta)(J\xi) &= \pi'(J\xi)J\eta = \lim_{n\to\infty} \pi'(J\xi)J\eta_n \\ &= \lim_{n\to\infty} J(\xi\eta_n) = \lim_{n\to\infty} J\pi(\xi)\eta_n \\ &= J\pi(\xi)\eta = J(\xi\eta) .\end{aligned}$$

Hence we have

$$\|(J\eta)(J\xi)\| \leq \|\pi'(\eta)\|\|\xi\| .$$

Therefore, recalling that $J\mathfrak{A}' \subset \mathfrak{D}^{\#}$, we can conclude that $J\eta$ belongs to $\mathfrak{A}''$. Thus we get $J\mathfrak{A}' = \mathfrak{A}''$. By symmetry, we also have $J\mathfrak{A}'' = \mathfrak{A}'$.

11. Tensor Product and Direct Sum of Modular Hilbert Algebras

Let $\mathfrak{A}_1$ and $\mathfrak{A}_2$ be two modular Hilbert algebras. Let $\mathfrak{H}_1$ and $\mathfrak{H}_2$ be their completions respectively. Let $\mathfrak{A}_1 \otimes \mathfrak{A}_2 = \mathfrak{A}$ be the algebraic tensor product of $\mathfrak{A}_1$ and $\mathfrak{A}_2$. Then $\mathfrak{A}_1 \otimes \mathfrak{A}_2$ has the natural involutive algebra structure defined by:

$$(\xi_1 \otimes \xi_2)(\eta_1 \otimes \eta_2) = \xi_1\eta_1 \otimes \xi_2\eta_2 ,$$

$$(\xi_1 \otimes \xi_2)^{\#} = \xi_1^{\#} \otimes \xi_2^{\#} .$$

Also $\mathfrak{A}_1 \otimes \mathfrak{A}_2$ has the inner product defined by:

$$(\xi_1 \otimes \xi_2 | \eta_1 \otimes \eta_2) = (\xi_1|\eta_1)(\xi_2|\eta_2) .$$

Let $\mathfrak{H}$ be the completion of $\mathfrak{A}$. Then $\mathfrak{H}$ is nothing but the tensor product $\mathfrak{H}_1 \otimes \mathfrak{H}_2$ of $\mathfrak{H}_1$ and $\mathfrak{H}_2$ as Hilbert space. Let $\Delta_1(\alpha)$ and $\Delta_2(\alpha)$ be the modular automorphisms of $\mathfrak{A}_1$ and $\mathfrak{A}_2$ respectively. Put $\Delta(\alpha) = \Delta_1(\alpha) \otimes \Delta_2(\alpha)$ for each complex number α. Then $\Delta(\alpha)$ is defined on $\mathfrak{A}_1 \otimes \mathfrak{A}_2$.

THEOREM 11.1. The tensor product $\mathfrak{A}_1 \otimes \mathfrak{A}_2$ of two modular Hilbert algebras $\mathfrak{A}_1$ and $\mathfrak{A}_2$ is a modular Hilbert algebra, and we hace

$$\mathfrak{L}(\mathfrak{A}_1 \otimes \mathfrak{A}_2) = \mathfrak{L}(\mathfrak{A}_1) \otimes \mathfrak{L}(\mathfrak{A}_2) ,$$

$$\mathfrak{L}(\mathfrak{A}_1 \otimes \mathfrak{A}_2)' = \mathfrak{L}(\mathfrak{A}_1)' \otimes \mathfrak{L}(\mathfrak{A}_2)' .$$

Proof. It is clear that $\{\mathfrak{A}_1 \otimes \mathfrak{A}_2, \Delta(\alpha)\}$ satisfies the postulates for modular Hilbert algebras except possibly postulate (VIII), so we shall only prove postulate (VIII). Let Δ_1 and Δ_2 be the modular operators of $\mathfrak{A}_1$ and $\mathfrak{A}_2$ respectively. If t is a real number, then $(1 + \Delta_i^t)^{-1}$ and $\Delta_i^t(1 + \Delta_i^t)^{-1}$ are both bounded positive operators on $\mathfrak{H}_i$, whose ranges are dense in $\mathfrak{H}_i$, $i = 1,2$. Put

$$W(t) = (1 + \Delta_1^t)^{-1} \otimes (1 + \Delta_2^t)^{-1} + \Delta_1^t(1 + \Delta_1^t)^{-1} \otimes \Delta_2^t(1 + \Delta_2^t)^{-1} .$$

Then $W(t)$ is a bounded positive operator on $\mathfrak{H}_1 \otimes \mathfrak{H}_2$. Since $W(t) \geq (1 + \Delta_1^t)^{-1} \otimes (1 + \Delta_2^t)^{-1}$, and $(1 + \Delta_1^t)^{-1} \otimes (1 + \Delta_2^t)^{-1}$ is a bounded positive operator on $\mathfrak{H}_1 \otimes \mathfrak{H}_2$, the range of $W(t)$ is also dense in $\mathfrak{H}_1 \otimes \mathfrak{H}_2$. By the equality:

$$W(t)((1 + \Delta_1^t)\mathfrak{A}_1 \otimes (1 + \Delta_2^t)\mathfrak{A}_2) = (1 + \Delta_1^t \otimes \Delta_2^t)(\mathfrak{A}_1 \otimes \mathfrak{A}_2) ,$$

$(1 + \Delta_1^t \otimes \Delta_2^t)(\mathfrak{A}_1 \otimes \mathfrak{A}_2)$ is dense in $\mathfrak{H}_1 \otimes \mathfrak{H}_2$. Therefore, $(1 + \Delta_1(t) \otimes \Delta_2(t))(\mathfrak{A}_1 \otimes \mathfrak{A}_2)$ is dense in $\mathfrak{H}_1 \otimes \mathfrak{H}_2$. Thus $\mathfrak{A}_1 \otimes \mathfrak{A}_2$ is a modular Hilbert algebra.

Since $\pi(\xi_1 \otimes \xi_2) = \pi(\xi_1) \otimes \pi(\xi_2)$, it is clear that

$$\mathfrak{L}(\mathfrak{A}_1 \otimes \mathfrak{A}_2) = \mathfrak{L}(\mathfrak{A}_1) \otimes \mathfrak{L}(\mathfrak{A}_2) .$$

Furthermore, the equality: $\pi'(\xi_1 \otimes \xi_2) = \pi'(\xi_1) \otimes \pi'(\xi_2)$, $\xi_1 \in \mathfrak{A}_1$, and $\xi_2 \in \mathfrak{A}_2$, implies that

$$\pi'(\mathfrak{A}_1 \otimes \mathfrak{A}_2)'' = \pi'(\mathfrak{A}_1)'' \otimes \pi'(\mathfrak{A}_2)'' .$$

Hence Theorem 4.1 assures that

$$\mathfrak{L}(\mathfrak{A}_1 \otimes \mathfrak{A}_2)' = \mathfrak{L}(\mathfrak{A}_1)' \otimes \mathfrak{L}(\mathfrak{A}_2)'.$$

This completes the proof.

Let $\{\mathfrak{A}_i\}_{i\in I}$ be a family of modular Hilbert algebras. Let $\Delta_i(\alpha)$ be a modular automorphism of each $\mathfrak{A}_i$, $i \in I$. Let $\mathfrak{A}$ denote the algebraic direct sum of $\{\mathfrak{A}_i\}_{i\in I}$. Then $\mathfrak{A}$ becomes naturally an involutive algebra and pre-Hilbert space. Let $\Delta(\alpha)$ be the algebraic direct sum of $\{\Delta_i(\alpha)\}_{i\in I}$ defined on $\mathfrak{A}$. Then it is clear that $\{\mathfrak{A},\Delta(\alpha)\}$ satsifies the postulates for modular Hilbert algebras except possibly postulate

(VIII). However, since for each real number t $(1 + \Delta(t))\mathfrak{A}$ is the algebraic direct sum of $\{(1 + \Delta_i(t))\mathfrak{A}_i\}_{i\in I}$ and each $(1 + \Delta_i(t))\mathfrak{A}_i$ is dense in $\mathfrak{A}_i$, $(1 + \Delta(t))\mathfrak{A}$ is dense in $\mathfrak{A}$; hence postulate (VIII) is also satisfied. Therefore we get the following:

THEOREM 11.2. If $\{\mathfrak{A}_i\}_{i\in I}$ is a family of modular Hilbert algebras, then the algebraic direct sum $\mathfrak{A}$ of $\{\mathfrak{A}_i\}_{i\in I}$ is also a modular Hilbert algebra with the left von Neumann algebra:

$$\mathfrak{L}(\mathfrak{A}) = \sum_{i\in I} \oplus \mathfrak{L}(\mathfrak{A}_i) .$$

12. The Standard Representation of Von Neumann Algebras

THEOREM 12.1. Let M be a von Neumann algebra acting on a Hilbert space $\mathfrak{H}$. If M admits a separating and generating vector ξ_0 in $\mathfrak{H}$, then there exists a modular Hilbert algebra $\mathfrak{B}$ such that M is spatially isomorphic to the left von Neumann algebra $\mathfrak{L}(\mathfrak{B})$ of $\mathfrak{B}$. Therefore, there exists an isometric involution J in $\mathfrak{H}$ such that

$$JMJ = M' \quad \text{and} \quad JM'J = M .$$

Proof. By Theorem 10.1, it is sufficient to show that there is a generalized Hilbert algebra structure in $M\xi_0 = \mathfrak{A}$ such that $\mathfrak{L}(\mathfrak{A}) = M$. For each $x \in M$, put

$$\xi_0(x) = x\xi_0 .$$

Define a product and an involution in $\mathfrak{A}$ by:

$$\xi_0(x)\xi_0(y) = \xi_0(xy),\ x,y \in M ;$$

$$\xi_0(x)^\# = \xi_0(x^*),\ x \in M .$$

Then it is clear that $\mathfrak{A}$ satisfies the postulates for generalized Hilbert algebras except possibly postulate (IX). However, for each $y \in M'$, we have

$$\begin{aligned}(\xi_0(x)^\# | y\xi_0) &= (x^*\xi_0 | y\xi_0) \\ &= (\xi_0 | xy\xi_0) = (\xi_0 | yx\xi_0) \\ &= (y^*\xi_0 | x\xi_0) = (y^*\xi_0 | \xi_0(x)) .\end{aligned}$$

Hence the involution $\xi_0(x) \mapsto \xi_0(x)^\#$ in $\mathfrak{A}$ admits the adjoint involution $y\xi_0 \mapsto y^*\xi_0$, $y \in M'$, with dense domain $M'\xi_0$. Therefore postulate (IX) holds. This completes the proof.

THEOREM 12.2. For every von Neumann algebra M, there exists a modular Hilbert algebra $\mathfrak{A}$ such that

$$M \cong \mathfrak{L}(\mathfrak{A}) .$$

Proof. If M is σ-finite, then M can be realized as a von Neumann algebra acting on a Hilbert with a separating and generating vector, because M has a faithful normal state which induces the desired representation of M. Hence Theorem 12.1 implies our assertion for M. In general, there exists an orthogonal family $\{e_i\}_{i\in I}$ of σ-finite projections and a family $\{\mathcal{B}_i\}_{i\in I}$ of type I factors such that $M \cong \Sigma_{i\in I} \oplus (e_iMe_i) \otimes \mathcal{B}_i$. Since each $\mathcal{B}_i$ is semi-finite, there is a Hilbert algebra $\mathfrak{B}_i$ such that $\mathfrak{L}(\mathfrak{B}_i) \cong \mathcal{B}_i$. By the remark above, there exists a modular Hilbert algebra $\mathfrak{A}_i$ such that $\mathfrak{L}(\mathfrak{A}_i) \cong e_i M e_i$ for each $i \in I$. Therefore, Theorems 11.1 and 11.2 imply that the algebraic direct sum $\mathfrak{A}$ of the algebraic tensor products $\mathfrak{A}_i \otimes \mathfrak{B}_i$ is the required modular Hilbert algebra.

THEOREM 12.3. Let M_1 and M_2 be two von Neumann algebras acting on Hilbert spaces $\mathfrak{H}_1$ and $\mathfrak{H}_2$ respectively. Let $M_1 \otimes M_2$ be the tensor product of M_1 and M_2 as von Neumann algebra, acting on the tensor product Hilbert space $\mathfrak{H}_1 \otimes \mathfrak{H}_2$. Then we have

$$(M_1 \otimes M_2)' = M_1' \otimes M_2' .$$

<u>Proof</u>. Let $\mathfrak{A}_1$ and $\mathfrak{A}_2$ be the modular Hilbert algebras with $\mathfrak{L}(\mathfrak{A}_1) \cong M_1$ and $\mathfrak{L}(\mathfrak{A}_2) \cong M_2$, whose existence is assured by Theorem 12.2. Then by Dixmier's Theorem [3: Theorem 3, p. 55] there exist Hilbert spaces $\mathfrak{K}_1$ and $\mathfrak{K}_2$ and projections $e_1' \in (\mathfrak{L}(\mathfrak{A}_1) \otimes 1_{\mathfrak{K}_2})'$ and $e_2' \in (\mathfrak{L}(\mathfrak{A}_2) \otimes 1_{\mathfrak{K}_2})'$ such that M_1 and M_2 are spatially isomorphic to $(\mathfrak{L}(\mathfrak{A}_1) \otimes 1_{\mathfrak{K}_1})_{e_1'}$ and $(\mathfrak{L}(\mathfrak{A}_2) \otimes 1_{\mathfrak{K}_2})_{e_2'}$ respectively. Hence $M_1 \otimes M_2$ is spatially isomorphic to $(\mathfrak{L}(\mathfrak{A}_1) \otimes \mathfrak{L}(\mathfrak{A}_2) \otimes 1_{\mathfrak{K}_1 \otimes \mathfrak{K}_2})_{e_1' \otimes e_2'}$ Therefore $(M_1 \otimes M_2)'$ is the image of $(\mathfrak{L}(\mathfrak{A}_1) \otimes \mathfrak{L}(\mathfrak{A}_2) \otimes 1_{\mathfrak{K}_1 \otimes \mathfrak{K}_2})'_{e_1' \otimes e_2'}$ by the spatial isomorphism. But we know

$$\begin{aligned}(\mathfrak{L}(\mathfrak{A}_1) &\otimes \mathfrak{L}(\mathfrak{A}_2) \otimes 1_{\mathfrak{K}_1 \otimes \mathfrak{K}_2})'_{e_1' \otimes e_2'} \\ &= [(\mathfrak{L}(\mathfrak{A}_1) \otimes \mathfrak{L}(\mathfrak{A}_2))' \otimes \mathfrak{B}(\mathfrak{K}_1 \otimes \mathfrak{K}_2)]_{e_1' \otimes e_2'} \\ &= (\mathfrak{L}(\mathfrak{A}_1)' \otimes \mathfrak{B}(\mathfrak{K}_1))_{e_1'} \otimes (\mathfrak{L}(\mathfrak{A}_2)' \otimes \mathfrak{B}(\mathfrak{K}_2))_{e_2'}\end{aligned}$$

which is isomorphic to $M_1' \otimes M_2'$ under the spatial isomorphism. This completes the proof.

This result is an affirmative answer to a problem which has remained unsettled for a long time, see for example [3: p. 28].

13. The Modular Automorphism Group and the Kubo-Martin-Schwinger Boundary Condition

Now, we are in the position to discuss the connection between Tomita's theory, which has been established above, and the Kubo-Martin-Schwinger (KMS) boundary condition which arises in the quantum statistical thermodynamics.

Let M be a von Neumann algebra. Suppose there exists a one parameter *-automorphism group σ_t, $-\infty < t < \infty$, of M, where the continuity of σ_t is considered as follows; for each $x \in M$, the function: $t \to \sigma_t(x)$ is strongly continuous. Remark that we cannot expect the uniform continuity of the function: $t \to \sigma_t(x)$, which is too strong as assumption.

DEFINITION 13.1. If a normal faithful positive linear functional φ_0 of M satsifies the following conditions, then it is said to satisfy the Kubo-Martin-Schwinger (KMS) boundary condition:

(13.1) φ_0 is invariant under σ_t ;

(13.2) There is a constant $\beta > 0$ and for each pair a, b in M, there exists a function F holomorphic in $0 < \mathrm{Im}\, z < \beta$ with boundary values:

$$F(t) = \varphi_0(\sigma_t(a)b)) \quad \text{and} \quad F(t + i\beta) = \varphi_0(b\sigma_t(a)) .$$ [2]

[2] Dr. Winnink pointed out to the author that (13.1) is derived from (13.2) by Sturm's Theorem.

Considering the one-parameter automorphism group $\sigma_{t/\beta}$, we may assume $\beta = 1$. If φ_0 is a trace of M, then σ_t must be the trivial automorphism group by Sturm's theorem, as shown by Hugenholtz [10].

THEOREM 13.1. Let M be a von Neumann algebra and φ_0 a normal faithful positive linear functional of M. Then there exists a one-parameter automorphism group σ_t of M such that φ_0 satisfies the KMS-boundary condition for $\beta = 1$.

Proof. Considering the cyclic representation of M induced by φ_0, we may assume that M acts on a Hilbert space $\mathfrak{H}$, which contains a separating and cyclic vector ξ_0 such that

$$\varphi_0(x) = (x\xi_0|\xi_0), \quad x \in M .$$

By Theorem 12.1, there is a modular Hilbert algebra $\mathfrak{B}$ in the generalized Hilbert algebra $\mathfrak{A} = M\xi_0$, with the left von Neumann algebra $\mathfrak{L}(\mathfrak{B}) = M$. From the construction of $\mathfrak{B}$ in Theorem 10.1, $\mathfrak{B}$ contains ξ_0 as a unit which is invariant under the involution J and the modular operator Δ. Define a one-parameter automorphism group σ_t of M by:

$$(13.3) \qquad \sigma_t(x) = \Delta^{it}x\Delta^{-it}, \quad x \in M, \quad -\infty < t < \infty .$$

Obviously the function: $t \to \sigma_t(x)$ is strongly continuous for every $x \in M$. Take a pair ξ, η in $\mathfrak{B}$. Then the function $\alpha \to (\Delta^{\alpha}\xi|\eta)$ is holomorphic on the whole plane C and satisfies the following equality:

$$(13.4) \qquad (\Delta^{\alpha}\xi|\eta^{\#}) = (\eta|\Delta^{1-\bar{\alpha}}\xi^{\#}), \quad \xi,\eta \in \mathfrak{B} .$$

If η is in $\mathfrak{A} = M\xi_0$, then there exists a sequence $\{\eta_n\}$ in $\mathfrak{B}$ such that $\eta = \lim \eta_n$ and $\eta^{\#} = \lim \eta_n^{\#}$. Then the functions $F_n(\alpha) = (\Delta^{\alpha}\xi|\eta_n^{\#})$

converge uniformly to $F(\alpha) = (\Delta^{\alpha}\xi|\eta^{\#})$ and also $(\eta_n|\Delta^{1-\bar{\alpha}}\xi^{\#})$ converge uniformly to $(\eta|\Delta^{1-\bar{\alpha}}\xi^{\#})$. Hence (13.4) remains true for every $\alpha \in C$ and every $\eta \in M\xi_0$ and $\xi \in \mathfrak{B}$. Now, take and fix a $\xi \in M\xi_0$. Since $\xi^{\#}$ is in $\mathfrak{D}(\Delta^{\frac{1}{2}}) = \mathfrak{D}^{\#}$ and $\mathfrak{B}$ is dense in the Hilbert space $\mathfrak{D}(\Delta^{\frac{1}{2}})$, we can find a sequence $\{\xi_n\}$ in $\mathfrak{B}$ with $\xi^{\#} = \lim_{n\to\infty} \xi_n^{\#}$ and $\Delta^{\frac{1}{2}}\xi^{\#} = \lim_{n\to\infty} \Delta^{\frac{1}{2}}\xi_n^{\#}$. Since $\Delta^t(1 + \Delta^{\frac{1}{2}})^{-1}$ is bounded for $0 \leq t \leq \frac{1}{2}$, we have

$$\Delta^t\xi^{\#} = \lim_{n\to\infty} \Delta^t\xi_n^{\#}, \quad 0 \leq t \leq \tfrac{1}{2} . \tag{13.5}$$

Since $\Delta^{\frac{1}{2}}\xi^{\#} = J\xi$, we have $\xi = \lim_{n\to\infty} \xi_n$ and

$$\Delta^{\frac{1}{2}}\xi = J\xi^{\#} = \lim J\xi_n^{\#} = \lim_{n\to\infty} \Delta^{\frac{1}{2}}\xi_n ,$$

so that we have

$$\Delta^t\xi = \lim_{n\to\infty} \Delta^t\xi_n, \quad 0 \leq t \leq \tfrac{1}{2} . \tag{13.5'}$$

Define a function G on $0 \leq \operatorname{Re} z \leq \frac{1}{2}$ by:

$$G(z) = (\Delta^z\xi|\eta^{\#}) .$$

Also define a function G' on $\frac{1}{2} \leq \operatorname{Re} z \leq 1$ by:

$$G'(z) = (\eta|\Delta^{1-\bar{z}}\xi^{\#}) .$$

Then by (13.5) and (13.5'), $G(z)$ and $G'(z)$ are both the limits of the following functions:

$$G_n(z) = (\Delta^z\xi_n|\eta^{\#}), \quad G_n'(z) = (\eta|\Delta^{1-\bar{z}}\xi_n^{\#}) .$$

Since $\|\Delta^z(1 + \Delta^{\frac{1}{2}})^{-1}\| \leq 1$ if $0 \leq \operatorname{Re} z \leq \frac{1}{2}$, the convergence of $G_n(z)$

and $G_n'(z)$ aro both uniform in the regions $0 \leq \mathrm{Re}\, z \leq \frac{1}{2}$ and $\frac{1}{2} \leq \mathrm{Re}\, z \leq 1$ respectively; hence $G(z)$ and $G'(z)$ are holomorphic in $0 < \mathrm{Re}\, z < \frac{1}{2}$ and $\frac{1}{2} < \mathrm{Re}\, z < 1$ respectively and $G(z) = G'(z)$ if $\mathrm{Re}\, z = \frac{1}{2}$ by (13.4). Hence $G(z)$ and $G'(z)$ define a function $H(z)$ holomorphic in $0 < \mathrm{Re}\, z < 1$ and continuous on the boundary. Put $F(z) = H(-iz)$. Then $F(z)$ is holomorphic in the strip $0 < \mathrm{Im}\, z < 1$ and continuous on the boundary.

Now take a pair x, y in M and put $\xi = x\xi_0$ and $\eta = y\xi_0$. Then we have

$$F(t) = G(-it) = (\Delta^{-it}\xi | \xi^{\#}) = (\xi | \Delta^{it}\eta^{\#})$$

$$= \varphi_0(\sigma_t(y)x)$$

and

$$F(t + i) = G'(-it + 1) = (\eta | \Delta^{-it}\xi^{\#})$$

$$= (\Delta^{it}\eta | \xi^{\#}) = \varphi_0(x\sigma_t(y)) \ .$$

Therefore, φ_0 satisfies condition (13.2) with respect to σ_t. This completes the proof.

The one-parameter automorphism group σ_t of M defined by (13.3) is called the modular automorphism group of M associated with φ_0. Then the modular automorphism group σ_t of M is uniquely determined by conditions (13.1) and (13.2) as seen in the following:

THEOREM 13.2. Let M be a von Neumann algebra. Let σ_t, $-\infty < t < +\infty$, be a one-parameter automorphism group of M. If a faithful normal positive linear functional φ_0 of M satisfies the

KMS-boundary condition for $\beta = 1$, then σ_t is the modular automorphism group of M associated with φ_0.[3)]

<u>Proof</u>. Let $\mathfrak{H}$, ξ_0 and $\mathfrak{A}$ be as in the proof of Theorem 13.1. Let $\mathcal{E}$ be the space defined in the proof of Theorem 10.1. Define a one-parameter unitary group $U(t)$ by

$$U(t)x\xi_0 = \sigma_t(x)\xi_0, \quad x \in M .$$

Then it is clear that $U(t)$ is a strongly continuous one-parameter unitary group which leaves ξ_0 invariant. By Stone's Theorem, there exists a self-adjoint operator H with dense domain $\mathfrak{D}(H)$ such that

$$U(t) = \exp itH, \quad H\xi = \lim_{t\to 0} \frac{1}{it}(U(t) - 1)\xi, \quad \xi \in \mathfrak{D}(H) . \tag{13.6}$$

For each function f in $\mathcal{E}$, $f(H)$ is bounded and $f(H)\mathfrak{A} \subset \mathfrak{A}$. In fact, we have

$$f(H)x\xi_0 = \left(\int_{-\infty}^{\infty} \hat{f}(s)\sigma_s(x)ds\right)\xi_0 ,$$

where $\hat{f}$ denotes the Fourier transform of f, which is integrable. Let $\mathfrak{B}$ denote the space of all $f(H)\xi$, $\xi \in \mathfrak{A}$, $f \in \mathcal{E}$. Then $\mathfrak{B}$ is dense in the definition domain of the function $f(H)$ of H for every continuous function f of a real variable. In particular, $\mathfrak{B}$ is dense in the domain of $\exp(H)$. Let e_α denote the function $e_\alpha(t) = e^{\alpha t}$. Then for each $f \in \mathcal{E}$, we have $e_\alpha f \in \mathcal{E}$ and

$$(e_\alpha f)(H) = \exp(\alpha H)f(H) .$$

Hence the function: $\alpha \to \exp(\alpha H)\xi$, $\xi \in \mathfrak{B}$, is analytic on the whole

---

3) The unicity of σ_t is also obtained by Winnink independently, which is mentioned in Cargèse lecture in theoretical physics (1969).

plane C. Take and fix a ξ in $\mathfrak{B}$. Then there is an $x \in M$ with $\xi = x\xi_0$. For each $\eta = y\xi_0 \in \mathfrak{A}$, put $F(t) = (U(-t)\xi|\eta^\#)$. Then we have

$$F(t) = (x\xi_0|U(t)\eta^\#) = (x\xi_0|\sigma_t(y^*)\xi_0)$$

$$= \varphi_0(\sigma_t(y)x) .$$

Hence by the KMS-boundary condition, we have

$$F(t + i) = \varphi_0(x\sigma_t(y)) .$$

On the other hand, the function: $\alpha \to (\exp(-i\alpha H)\xi|\eta)$ is analytic on the whole plane C and coincides with $F(t)$ on the real line. Hence we have

$$\varphi_0(x\sigma_t(y)) = (\exp(-i(t + i)H)\xi|\eta^\#)$$

$$= (\exp(H)\exp(-itH)\xi|\eta^\#)$$

On the other hand, we have

$$\varphi_0(x\sigma_t(y)) = (xU(t)y\xi_0|\xi_0) = (\exp(itH)\eta|\xi^\#) .$$

Hence we get

$$(\exp(H)\xi|\exp(itH)\eta^\#) = (\exp(itH)\eta|\xi^\#) .$$

Replacing η by $\exp(-itH)\eta$, we get

$$(\exp(H)\xi|\eta^\#) = (\eta|\xi^\#), \quad \eta \in \mathfrak{A} .$$

Therefore $\exp(H)\xi$ belongs to $\mathcal{D}^\flat$ and we have

$$(\exp(H)\xi)^\flat = \xi^\#, \quad \xi \in \mathfrak{B} ,$$

it follows that

(13.7) $$\exp(H)\xi = \Delta\xi, \quad \xi \in \mathfrak{B} .$$

Since $\exp(H)$ is the closure of its restriction to $\mathfrak{B}$, $\exp(H)$ is contained in Δ; hence by the maximality of self-adjoint operators, we have

$$(13.8) \qquad \Delta = \exp H .$$

Therefore, the one parameter group $U(t)$ coincides with Δ^{it}, which means that the originally given automorphism group σ_t is just the modular automorphism group of M associated with φ_0. This completes the proof.

The KMS-boundary condition is also defined for states of C^*-algebra if a one parameter automorphism group is given. Let A be a C^*-algebra. Suppose a one parameter automorphism group σ_t of A is given. If a state φ of A satisfies conditions (13.1) and (13.2), where we have to assume the continuity of F in (13.2) on the boundary because we do not assume the continuity of the one parameter group σ_t; then φ is said to satisfy the KMS-boundary condition.[4] Then we have the following result concerning the KMS-boundary conditions based on von Neumann algebras and those based on C^*-algebras.

THEOREM 13.3. Let A be a C^*-algebra which admits a one parameter automorphism group σ_t. Suppose φ is a state of A satisfying the KMS-boundary condition with respect to σ_t. Then we have the following:

1° The left kernel $\mathfrak{m}_\varphi = \{x \in A; \varphi(x^*x) = 0\}$ is a two-sided ideal of A, and coincides with the kernel $\pi_\varphi^{-1}(0)$ of the cyclic representation π_φ of A induced by φ;

2° There exists a unique faithful normal state $\tilde{\varphi}$ of the

---

[4] The continuity assumption for F on the boundary is not necessary, because the measurability of F on the boundary implies the strong

von Neumann algebra M generated by $\pi_\varphi(A)$ such that

(i) $\varphi = {}^t\pi_\varphi(\tilde{\varphi})$,

(ii) π_φ transforms σ_t onto the modular automorphism $\tilde{\sigma}_t$ of M associated with $\tilde{\varphi}$, that is,

$$\tilde{\sigma}_t \cdot \pi_\varphi(x) = \pi_\varphi \cdot \sigma_t(x), \quad x \in A .$$

<u>Proof</u>. We shall assume $\beta = 1$, if necessary, changing the scale by t/β. If x is in $\mathfrak{m}_\varphi$, then $\varphi(y^*x) = 0$ for every $y \in A$ by the Schwarz's inequality. Let F be an analytic function of complex variable with boundary value:

$$F(t) = \varphi(\sigma_t(y^*)x) \quad \text{and} \quad F(t + i) = \varphi(x\sigma_t(y^*)) .$$

Since $F(t) = 0$ for all t, F vanishes identically; hence we have $\varphi(x\sigma_t(y^*)) = 0$. In particular, $\varphi(xx^*) = 0$. Therefore, $\mathfrak{m}_\varphi$ coincides with the right kernel $\{x \in A = \varphi(xx^*) = 0\}$, which means that $\mathfrak{m}_\varphi$ is a two-sided ideal of A. For $a \in A$ to be in $\pi_\varphi(0)$, it is necessary and sufficient that ax is in $\mathfrak{m}_\varphi$ for every $x \in A$. Since $\mathfrak{m}_\varphi$ is the right ideal of φ, that ax is in $\mathfrak{m}_\varphi$ is equivalent to that $\varphi((ax)(ax)^*) = 0$. Hence $a \in A$ is in $\pi_\varphi^{-1}(0)$ if and only if $\varphi(aa^*) = 0$. Thus $\pi_\varphi^{-1}(0)$ coincides with $\mathfrak{m}_\varphi$.

Let ξ_0 be the cyclic vector of π_φ corresponding to φ. Let $\tilde{\varphi}$ be the cyclic state of $M = \pi_\varphi(A)''$ defined by $\tilde{\varphi}(x) = (x\xi_0|\xi_0)$, $x \in M$. Noticing that σ_t leaves the ideal $\mathfrak{m}_\varphi = \pi_\varphi^{-1}(0)$ invariant because of the invariance of the state φ under σ_t, we define a one parameter automorphism $\tilde{\sigma}_t$ of $\pi_\varphi(A)$ by

$$\tilde{\sigma}_t \circ \pi_\varphi(x) = \pi_\varphi \circ \sigma_t(x), \quad x \in A .$$

Then $\tilde{\sigma}_t$ is induced by the unitary operators $U(t)$, which is defined by $U(t)\pi_\varphi(x)\xi_0 = \pi_\varphi \circ \sigma_t(x)\xi_0$, $x \in A$. Hence $\tilde{\sigma}_t$ can be extended to a one parameter automorphism group of M, which is also denoted by $\tilde{\sigma}_t$. By the continuity of the function:

$$t \in R \to (U(t)\pi_\varphi(x)\xi_0 \,|\, \pi_\varphi(y)\xi_0) = \varphi(y^*\sigma_t(x)), \quad x,y \in A ,$$

$U(t)$ is a strongly continuous one parameter unitary group. Hence $t \mapsto \tilde{\sigma}_t(x)$, $x \in M$, is strongly continuous. If x and y are in M, then there exist two bounded sequence $\{x_n\}$ and $\{y_n\}$ in $\pi(A)$ such that

$$x\xi_0 = \lim_{n\to\infty} x_n\xi_0, \quad x^*\xi_0 = \lim x_n^*\xi_0 ;$$

$$y\xi_0 = \lim y_n\xi_0, \quad y^*\xi_0 = \lim y_n^*\xi_0 .$$

Then we have

$$\tilde{\varphi}(\tilde{\sigma}_t(x)y) = (y\xi_0 \,|\, U(t)x^*\xi_0)$$

$$= \lim(y_n\xi_0 \,|\, U(t)x_n^*\xi_0) ;$$

$$\tilde{\varphi}(y\tilde{\sigma}_t(x)) = (U(t)x\xi_0 \,|\, y^*\xi_0)$$

$$= \lim(U(t)x_n\xi_0 \,|\, y_n^*\xi_0) .$$

For each $n = 1,2,\ldots$, let F_n be a function of complex variable holomorphic in the stripe $0 < \operatorname{Im} z < 1$ with the boundary value:

$$F_n(t) = \tilde{\varphi}(\tilde{\sigma}_t(x_n)y_n) = (y_n\xi_0 | U(t)x_n^*\xi_0) \ ;$$

$$F_n(t + i) = \tilde{\varphi}(y_n\tilde{\sigma}_t(x_n)) = (U(t)x_n\xi_0 | y_n\xi_0) \ .$$

Then $\{F_n(t)\}$ and $\{F_n(t + i)\}$ are both uniformly bounded with respect to t and converge uniformly. Therefore, by maximal modulas principle, $\{F_n(z)\}$ converges uniformly to a function $F(z)$ which is holomorphic in the stripe $0 < \mathrm{Im}\, z < 1$ and continuous on the stripe $0 \leq \mathrm{Im}\, z \leq 1$. Hence

$$F(t) = \tilde{\varphi}(\tilde{\sigma}_t(x)y) \quad \text{and} \quad F(t + i) = \tilde{\varphi}(y\tilde{\sigma}_t(x)) \ .$$

Therefore, $\tilde{\varphi}$ satisfies the KMS boundary condition with respect to $\tilde{\sigma}_t$. Thus $\tilde{\sigma}_t$ is the modular automorphism group of $\tilde{\varphi}$, where by the first assertion of our theorem for M, $\tilde{\varphi}$ and $\tilde{\sigma}_t$, $\tilde{\varphi}$ is faithful. This completes the proof.

The last half of the theorem was also mentioned by H. Araki in [1] in a little different form.

14. Semi-Finiteness and the Modular Automorphism Group

Let $\mathfrak{A}$ be a modular Hilbert algebra. We shall give a criterion for the semi-finiteness of the left von Neumann algebra $\mathfrak{L}(\mathfrak{A})$ in terms of the modular automorphism group.

As seen already in §10, the one parameter unitary group Δ^{it}, $-\infty < t < +\infty$, induces a one-parameter automorphism group σ_t of $\mathfrak{L}(\mathfrak{A})$ defined by:

$$(14.1) \qquad \sigma_t(x) = \Delta^{it} x \Delta^{-it}, \quad x \in \mathfrak{L}(\mathfrak{A}), \quad -\infty < t < +\infty .$$

The automoprhism group σ_t is called the <u>modular</u> automorphism group of $\mathfrak{L}(\mathfrak{A})$ associated with the modular Hilbert algebra $\mathfrak{A}$. Remark that σ_t is not determined a priori by the left von Neumann algebra $\mathfrak{L}(\mathfrak{A})$ itself unless we specify the modular Hilbert algebra $\mathfrak{A}$.

THEOREM 14.1. Suppose $\mathfrak{A}$ is a modular Hilbert algebra. If the modular automorphism group σ_t of the left von Neumann algebra $\mathfrak{L}(\mathfrak{A})$ of $\mathfrak{A}$ is inner, that is, there exists a one parameter unitary group $\Gamma(t)$ in $\mathfrak{L}(\mathfrak{A})$ which implements the automorphism group σ_t, then $\mathfrak{L}(\mathfrak{A})$ is semi-finite.

We shall divide the proof of the theorem into several steps and employ the arguments developed by J. Dixmier in [4, 5, p. 287-292], in a slightly modified form.

Putting $\Gamma'(t) = \Gamma(t)^{-1}\Delta^{it}$, $-\infty < t < \infty$, we get another one-parameter unitary group $\Gamma'(t)$, which belongs to $\mathfrak{L}(\mathfrak{A})'$ and

$$(14.2) \qquad \Delta^{it} = \Gamma(t)\Gamma'(t), \quad -\infty < t < +\infty .$$

By Stone's Theorem, $\Gamma(t)$ and $\Gamma'(t)$ can be written in the form:

$$\Gamma(t) = \exp(itH), \quad H = \lim_{t\to 0} \frac{1}{it} (\Gamma(t) - 1) ;$$

(14.3)

$$\Gamma'(t) = \exp(itH'), \quad H' = \lim_{t\to 0} \frac{1}{it} (\Gamma'(t) - 1) ,$$

where H and H' are self-adjoint and affiliated with $\mathfrak{L}(\mathfrak{A})$ and $\mathfrak{L}(\mathfrak{A})'$ respectively. Let

(14.4) $$H = \int_{-\infty}^{\infty} \lambda dF(\lambda), \quad H' = \int_{-\infty}^{\infty} \lambda dF'(\lambda)$$

be the spectral decompositions of H and H' respectively. Define projections $\{F_n\} \subset \mathfrak{L}(\mathfrak{A})$ and $\{F_n'\} \subset \mathfrak{L}(\mathfrak{A})'$ by:

$$F_n = \int_{-n}^{n} dF(\lambda), \quad F_n' = \int_{-n}^{n} dF'(\lambda) .$$

Thdn $\Gamma(t)\Gamma'(t)F_nF_m'$ becomes a uniformly continuous one parameter unitary group on the space $F_nF_m'\mathfrak{H}$, and we have

$$\Delta^{it}F_nF_m' = \Gamma(t)F_n\Gamma'(t)F_m' .$$

Therefore, we get

(14.5) $$\Delta^{i\alpha}F_nF_m' = \Gamma(\alpha)F_n\Gamma'(\alpha)F_m'$$

for every complex number α. In particular, we have

$$(14.5')\qquad \begin{cases} \Delta^{\frac{1}{2}}F_n F'_m = \Gamma(-\tfrac{1}{2}i)F_n\Gamma'(-\tfrac{1}{2}i)F'_m\ ; \\ \\ \Delta^{-\frac{1}{2}}F_n F'_m = \Gamma(\tfrac{1}{2}i)F_n\Gamma'(\tfrac{1}{2}i)F'_m\ , \end{cases}$$

which means that

$$(14.6)\qquad \begin{cases} \Delta^{\frac{1}{2}}F_n F'_m = \exp(\tfrac{1}{2}H)F_n\ \exp(\tfrac{1}{2}H')\ F'_m\ ; \\ \\ \Delta^{-\frac{1}{2}}F_n F'_m = \exp(-\tfrac{1}{2}H)F_n\ \exp(-\tfrac{1}{2}H')F'_m\ . \end{cases}$$

For $\xi \in \mathfrak{H}$ to be in the definition domain $\mathfrak{D}(\exp(\frac{1}{2}H'))$ it is necessary and sufficient that the sequence $\{\exp(\frac{1}{2}H')F'_m\xi\}_{m=1}^{\infty}$ converges strongly to some element η in $\mathfrak{H}$. If this is the case, then $\eta = \exp(\frac{1}{2}H')\,\xi$. Since $\exp(\frac{1}{2}H)F_n$ is bounded and $\Delta^{\frac{1}{2}}$ is closed, we have

$$\Delta^{\frac{1}{2}}F_n\xi = \exp(\tfrac{1}{2}H')\exp(\tfrac{1}{2}H)F_n\xi$$

for every $\xi \in \mathfrak{D}(\exp(\frac{1}{2}H')\exp(\frac{1}{2}H)F_n)$; hence

$$\Delta^{\frac{1}{2}}F_n \supset \exp(\tfrac{1}{2}H')\exp(\tfrac{1}{2}H)F_n\ .$$

On the other hand, since $\exp(\frac{1}{2}H)F_n$ belongs to $\mathfrak{L}(\mathfrak{A})$ and $\exp(\frac{1}{2}H')$ and $\exp(\frac{1}{2}H)F_n$ are both self-adjoint, $\exp(\frac{1}{2}H')\exp(\frac{1}{2}H)F_n$ is also self-adjoint. It is clear that $\Delta^{\frac{1}{2}}F_n$ is symmetric. Hence by the maximality of slef-adjoint operators we have

$$(14.7)\qquad \Delta^{\frac{1}{2}}F_n = \exp(\tfrac{1}{2}H')\exp(\tfrac{1}{2}H)F_n,\quad n = 1,2,\ldots\ .$$

Definition 14.1. If $\xi \in \mathfrak{H}$ satisfies the inequality

$$\sup\{\|\xi\eta\|;\ \eta \in \mathfrak{A}',\ \|\eta\| \leq 1\} < +\infty ,$$

then it is said to be left bounded.

If $\xi \in \mathfrak{H}$ is left bounded, then ξ induces a bounded operator $\pi(\xi)$ defined by $\pi(\xi)\eta = \xi\eta$, $\eta \in \mathfrak{A}'$. Since $\pi'(\eta)$, $\eta \in \mathfrak{A}'$, generates $\mathfrak{L}(\mathfrak{A})'$ by Theorem 3.1, $\pi(\xi)$ belongs to $\mathfrak{L}(\mathfrak{A})$. Remark that the left boundedness of $\xi \in \mathfrak{H}$ does not necessarily imply that ξ is in $\mathfrak{A}''$.

Let $\mathfrak{M}$ denote the set of all $\xi \in \mathfrak{D}(\exp(\frac{1}{2}H'))$ which are left bounded. Put

$$(14.8) \qquad \mathfrak{m} = \left\{ \sum_{i=1}^{n} \pi(\xi_i)\pi(\eta_i)^*;\ \xi_i,\eta_i \in \mathfrak{M},\ i = 1,2,\ldots,n \right\} .$$

LEMMA 14.1. If ξ is left bounded and x is in $\mathfrak{L}(\mathfrak{A})$, then $x\xi$ is also left bounded and

$$(14.9) \qquad \pi(x\xi) = x\pi(\xi) .$$

Hence $\mathfrak{m}$ is two sided ideal of $\mathfrak{L}(\mathfrak{A})$.

Proof. For each $\eta \in \mathfrak{A}'$, we have

$$(x\xi)\eta = \pi'(\eta)x\xi = x\pi'(\eta)\xi = x\pi(\xi)\eta ,$$

which implies immediately the first assertion. The second assertion follows from the calculation:

$$x \sum_{i=1}^{n} \pi(\xi_i)\pi(\eta_i)^* = \sum_{i=1}^{n} \pi(x\xi_i)\pi(\eta_i)^* ;$$

$$\left(\sum_{i=1}^{n} \pi(\xi_i)\pi(\eta_i)^* \right) x = \sum_{i=1}^{n} \pi(\xi_i)(x^*\pi(\eta_i))^*$$

$$= \sum_{i=1}^{n} \pi(\xi_i)\pi(x^*\eta_i)^* ,$$

because every element of $\mathfrak{L}(\mathfrak{A})$ leaves $\mathfrak{D}(\exp(\frac{1}{2}H'))$ invariant.

If $x\xi = 0$ for every $x \in \mathfrak{m}$, then $\pi(\eta)^*\xi = 0$ for every $\eta \in \mathfrak{M}$; hence for each $\zeta \in \mathfrak{A}'$ we have

$$(\xi|\pi'(\zeta)\eta) = (\xi|\pi(\eta)\zeta) = (\pi(\eta)^*\xi|\zeta) = 0 .$$

Since $\pi'(\mathfrak{A}')$ is not degenerate we have $\xi \in \mathfrak{M}^\perp$. Every element of $F_n\mathfrak{A}$ is left bounded and by (14.7) $F_n\mathfrak{A} \subset \mathfrak{D}(\exp(\frac{1}{2}H'))$, so that $(\xi|F_n\zeta) = 0$ for every $\zeta \in \mathfrak{A}$. Therefore $F_n\xi = 0$ for $n = 1,2,\ldots$; hence $\xi = 0$. Therefore, $\mathfrak{m}$ is non-degenerate, which implies that $\mathfrak{m}$ is σ-weakly dense in $\mathfrak{L}(\mathfrak{A})$.

LEMMA 14.2. If $\{\xi_i,\eta_i;\ i = 1,2,\ldots,n\}$ are contained in $\mathfrak{M}$, then the inequality $\Sigma_{i=1}^{n} \pi(\xi_i)^*\pi(\eta_i) \geq 0$ implies the inequality:

$$(14.10) \qquad \sum_{i=1}^{n} (J\exp(-\tfrac{1}{2}H)J\eta_i\,|\,J\exp(-\tfrac{1}{2}H)J\xi_i) \geq 0 ;$$

whenever $J\xi_i$ and $J\eta_i$, $i = 1,2,\ldots,n$ are in $\mathfrak{D}(\exp(-\frac{1}{2}H))$.

<u>Proof</u>. For each $\zeta \in \mathfrak{A}'$, we have

$$0 \leq \left(\sum_{i=1}^{n} \pi(\xi_i)^* \pi(\eta_i)\zeta | \zeta \right) = \sum_{i=1}^{n} (\pi(\eta_i)\zeta | \pi(\xi_i)\zeta)$$

$$= \sum_{i=1}^{n} (\pi'(\zeta)\eta_i | \pi'(\zeta)\xi_i) ;$$

hence for every $x \in \mathfrak{L}(\mathfrak{A})'$ we have

$$\sum_{i=1}^{n} (x\eta_i | x\xi_i) \geq 0 .$$

Since $J \exp(-\frac{1}{2}H)J$ is affiliated with $\mathfrak{L}(\mathfrak{A})'$ by Theorem 4.1, we can choose a net $\{x_\alpha\}$ in $\mathfrak{L}(\mathfrak{A})'$ such that

$$\lim_{\alpha} \sum_{i=1}^{n} (x_\alpha\eta_i | x_\alpha\xi_i) = \sum_{i=1}^{n} (J \exp(-\tfrac{1}{2}H)J\eta_i | J \exp(-\tfrac{1}{2}H)J\xi_i) ,$$

then our assertion follows.

LEMMA 14.3. If $\Sigma_{i=1}^{n} \pi(\xi_i)\pi(\eta_i)^* \in \mathfrak{M}$ is positive, then

$$\sum_{i=1}^{n} (\exp(\tfrac{1}{2}H')\xi_i | \exp(\tfrac{1}{2}H')\eta_i) \geq 0 . \tag{14.11}$$

Proof. By (14.7), $F_n\xi_i$ and $F_n\eta_j$ are all in $\mathfrak{D}(\Delta^{\frac{1}{2}}) = \mathfrak{D}^\#$, so that they are all in $\mathfrak{A}''$.

Noticing that

$$\sum_{i=1}^{n} \pi((F_n\xi_i)^\#)^*\pi((F_n\eta_i)^\#) = \sum_{i=1}^{n} \pi(F_n\xi_i)\pi(F_n\eta_i)^*$$

$$= F_n \left(\sum_{i=1}^{n} \pi(\xi_i)\pi(\eta_i)^* \right) F_n \geq 0 ,$$

we have, by Lemma 14.2 and (14.7),

$$\sum_{i=1}^{n} (\exp(\tfrac{1}{2}H')\, F_m\xi_i \,|\exp(\tfrac{1}{2}H')\, F_m\eta_i)$$

$$= \sum_{i=1}^{n} (\exp(-\tfrac{1}{2}H)\, \Delta^{\frac{1}{2}}F_m\xi_i \,|\exp(-\tfrac{1}{2}H)\, \Delta^{\frac{1}{2}}F_m\eta_i)$$

$$= \sum_{i=1}^{n} (J \exp(-\tfrac{1}{2}H)\, J(F_m\eta_i)^{\#} \,|J \exp(-\tfrac{1}{2}H) J(F_m\xi_i)^{\#})$$

$$\geq 0 \;;$$

hence

$$\sum_{i=1}^{n} (\exp(\tfrac{1}{2}H')\, \xi_i \,|\exp(\tfrac{1}{2}H')\, \eta_i) = \lim_{m\to\infty} \sum_{i=1}^{n} (F_m \exp(\tfrac{1}{2}H')\, \xi_i \,|F_m \exp(\tfrac{1}{2}H')\eta_i)$$

$$= \lim_{m\to\infty} \sum_{i=1}^{n} (\exp(\tfrac{1}{2}H')F_m\xi_i \,|\exp(\tfrac{1}{2}H')F_m\eta_i) \geq 0 \;,$$

which completes the proof.

Now we can define a functional τ on $\mathfrak{m}$ by:

$$(14.12) \qquad \tau\left(\sum_{i=1}^{n} \pi(\xi_i)\pi(\eta_i)^*\right) = \sum_{i=1}^{n} (\exp(\tfrac{1}{2}H')\xi_i \,|\exp(\tfrac{1}{2}H')\eta_i)$$

for every $\Sigma_{i=1}^{n} \pi(\xi_i)\pi(\eta_i)^* \in \mathfrak{m}$. By Lemma 14.3, the definition of τ does not involve any ambiguity. That is, if x in $\mathfrak{m}$ has two expressions:

$$x = \sum_{i=1}^{n} \pi(\xi_i)\pi(\eta_i)^* = \sum_{j=1}^{m} \pi(\xi_j')\pi(\eta_j')^* ,$$

then we have

$$\sum_{i=1}^{n} (\exp(\tfrac{1}{2}H')\,\xi_i \,|\exp(\tfrac{1}{2}H')\,\eta_i) = \sum_{j=1}^{m} (\exp(\tfrac{1}{2}H')\,\xi_j' \,|\exp(\tfrac{1}{2}H')\,\eta_j') .$$

Lemma 14.3 means also that τ is a positive linear functional on the ideal $\mathfrak{m}$ of $\mathcal{L}(\mathfrak{A})$. Take an x in $\mathcal{L}(\mathfrak{A})$. Then we have

$$\tau\left(x \sum_{i=1}^{n} \pi(\xi_i)\pi(\eta_i)^*\right) = \tau\left(\sum_{i=1}^{n} \pi(x\xi_i)\pi(\eta_i)^*\right)$$

$$= \sum_{i=1}^{n} (\exp(\tfrac{1}{2}H')\,x\xi_i \,|\exp(\tfrac{1}{2}H')\,\eta_i)$$

$$= \sum_{i=1}^{n} (x\,\exp(\tfrac{1}{2}H')\,\xi_i \,|\exp(\tfrac{1}{2}H')\,\eta_i)$$

$$= \sum_{i=1}^{n} (\exp(\tfrac{1}{2}H')\,\xi_i \,|x^*\,\exp(\tfrac{1}{2}H')\,\eta_i)$$

$$= \sum_{i=1}^{n} (\exp(\tfrac{1}{2}H')\,\xi_i \,|\exp(\tfrac{1}{2}H')\,x^*\eta_i)$$

$$= \tau\left(\sum_{i=1}^{n} \pi(\xi_i)\pi(x^*\eta_i)^*\right) = \tau\left(\left(\sum_{i=1}^{n} \pi(\xi_i)\pi(\eta_i)^*\right)x\right)$$

it follows that τ is central. Therefore, τ is a trace defined on the ideal $\mathfrak{m}$ of $\mathcal{L}(\mathfrak{A})$. By the density of $\mathfrak{m}$, τ is semi-finite.

LEMMA 14.4. Every positive operator in $\mathfrak{m}$ is of the form

$\Sigma_{i=1}^{n} \pi(\xi_i)\pi(\xi_i)^*$, $\xi_i \in \mathfrak{M}$.

Proof. Suppose $x = \Sigma_{i=1}^{n} \pi(\xi_i)\pi(\eta_i)^* \in \mathfrak{m}$ is positive. Put $\zeta_i = \frac{1}{2}(\xi_i + \eta_i)$ and $\zeta_i' = \frac{1}{2}(\xi_i - \eta_i)$. Then we

$$x = \sum_{i=1}^{n} (\pi(\zeta_i) + \pi(\zeta_i'))(\pi(\zeta_i) - \pi(\zeta_i'))^*$$

$$= \sum_{i=1}^{n} (\pi(\zeta_i)\pi(\zeta_i)^* - \pi(\zeta_i)\pi(\zeta_i')^* + \pi(\zeta_i')\pi(\zeta_i)^* - \pi(\zeta_i')\pi(\zeta_i')^*) ;$$

$$x^* = \sum_{i=1}^{n} (\pi(\zeta_i)\pi(\zeta_i)^* + \pi(\zeta_i)\pi(\zeta_i')^* - \pi(\zeta_i')\pi(\zeta_i)^* + \pi(\zeta_i')\pi(\zeta_i')^*)$$

Hence $x = \frac{1}{2}(x + x^*) = \Sigma_{i=1}^{n} \pi(\zeta_i)\pi(\zeta_i)^* - \Sigma_{i=1}^{n} \pi(\zeta_i')\pi(\zeta_i')^* \geq 0$. Put $y = \Sigma_{i=1}^{n} \pi(\zeta_i)\pi(\zeta_i)^*$. Then $y \geq x$; hence it follows by [3: Lemma 2, p. 11] that there is an element $u \in \mathfrak{L}(\mathfrak{A})$ such that $x^{\frac{1}{2}} = uy^{\frac{1}{2}}$ and $\|u\| \leq 1$. Then we get

$$x = uyu^* = \sum_{i=1}^{n} \pi(u\zeta_i)\pi(u\zeta_i)^*$$

as desired. This completes the proof.

As an immediate consequence of Lemma 14.4, it follows that τ is faithful because the equality

$$0 = \tau\left(\sum_{i=1}^{n} \pi(\xi_i)\pi(\xi_i)^*\right) = \sum_{i=1}^{n} \|\exp(\tfrac{1}{2} H')\xi_i\|^2$$

implies $\xi_i = 0$, $i = 1,2,\ldots,n$.

LEMMA 14.5. If ξ_i, $i = 1,\dots,n$, is in $\mathfrak{M}$, then there exists a $\xi \in \mathfrak{A}''$ with $\Sigma_{i=1}^n \pi(\xi_i)^*\pi(\xi_i) = \pi(\xi)^2$ and $\pi(\xi) \geq 0$.

Proof. Put $x = \Sigma_{i=1}^n \pi(\xi_i)^*\pi(\xi_i)$. Since $\|\pi(\xi_i)\eta\|^2 \leq \|x^{\frac{1}{2}}\eta\|^2$ for every $\eta \in \mathfrak{H}$, there exists $y_i \in \mathfrak{L}(\mathfrak{A})$ with $\pi(\xi_i) = y_i x^{\frac{1}{2}}$ and $y_i[x^{\frac{1}{2}}\mathfrak{H}]^{\perp} = 0$ for $i = 1,2,\dots,n$. For each $\eta \in \mathfrak{H}$, we have

$$\left(\sum_{i=1}^n y_i^* y_i x^{\frac{1}{2}}\eta \,|\, x^{\frac{1}{2}}\eta\right) = \left(\sum_{i=1}^n x^{\frac{1}{2}} y_i^* y_i x^{\frac{1}{2}}\eta \,|\, \eta\right)$$

$$= \left(\sum_{i=1}^n \pi(\xi_i)^*\pi(\xi_i)\eta \,|\, \eta\right) = \|x^{\frac{1}{2}}\eta\|^2 \ ;$$

hence $(\Sigma_{i=1}^n y_i^* y_i \eta \,|\, \eta) = \|\eta\|^2$ for each $\eta \in [x\mathfrak{H}]$ and $\Sigma_{i=1}^n y_i^* y_i [x\mathfrak{H}]^{\perp} = 0$, so that $\Sigma_{i=1}^n y_i^* y_i$ is the projection onto $[x\mathfrak{H}]$. Therefore we have

$$x^{\frac{1}{2}} = \left(\sum_{i=1}^n y_i^* y_i\right) x^{\frac{1}{2}} = \sum_{i=1}^n y_i^* \pi(\xi_i) = \sum_{i=1}^n \pi(y_i^* \xi_i)$$

$$= \pi\left(\sum_{i=1}^n y_i^* \xi_i\right),$$

so that $\xi = \Sigma_{i=1}^n y_i^* \xi_i$ is the required element in $\mathfrak{A}''$ because the self-adjointness of $\pi(\xi)$ implies that ξ is in $\mathfrak{D}^{\#}$.

Now we shall prove the normality of τ. Let $\{x_i\}_{i \in I}$ be an increasing net in $\mathfrak{m}^+$, the positive part of $\mathfrak{m}$, with $x = \sup x_i \in \mathfrak{m}^+$. Since it is clear that $\tau(x) \geq \sup \tau(x_i)$, we have to prove only $\tau(x) \leq \sup \tau(x_i)$. Take an $\varepsilon > 0$. Let x be $x = \Sigma_{i=1}^k \pi(\xi_i)\pi(\xi_i)^*$. Then we have

$$F_nxF_n = \sum_{i=1}^{k} \pi(F_n\xi_i)\pi(F_n\xi_i)^* ;$$

$$\tau(x) = \sum_{i=1}^{k} \|\exp(\tfrac{1}{2}H')\xi_i\|^2 ;$$

$$\tau(F_nxF_n) = \sum_{i=1}^{k} \|\exp(\tfrac{1}{2}H')F_n\xi_i\|^2 = \sum_{i=1}^{k} \|F_n \exp(\tfrac{1}{2}H')\xi_i\|^2 .$$

Hence we can choose an n with

$$\tau(x) - \varepsilon/2 \leq \tau(F_nxF_n) \leq \tau(x) .$$

We shall show that there is an $i \in I$ with

$$\tau(F_nxF_n) - \varepsilon/2 \leq \tau(F_nx_iF_n) \leq \tau(x_i) ,$$

which implies the desired inequality. Since $F_n\mathfrak{M} \subset \mathfrak{A}''$ by (14.7), F_nxF_n and $F_nx_iF_n$ by Lemma 14.4 and 14.5 can be written in the form:

$$F_nxF_n = \pi(\eta)^2, \quad \pi(\eta) \geq 0, \quad \eta \in \mathfrak{A}'' ;$$

$$F_nx_iF_n = \pi(\eta_i)^2, \quad \pi(\eta_i) \geq 0, \quad \eta_i \in \mathfrak{A}'' ;$$

by the reasoning in the first paragraph of the proof of Lemma 14.3. The equality $F_n\pi(\eta) = \pi(\eta)$ implies $F_n\eta = \eta$. Then equality (14.6) implies that η is in $\mathfrak{D}(\exp(\frac{1}{2}H'))$. Similarly η_i is in $\mathfrak{D}(\exp(\frac{1}{2}H'))$. Then

$$\tau(F_nxF_n) = \|\exp(\tfrac{1}{2}H')\eta\|^2 ;$$

$$\tau(F_nx_iF_n) = \|\exp(\tfrac{1}{2}H')\eta_i\|^2 .$$

Since $F_n x_i F_n$ converges strongly to $F_n x F_n$, so does $\pi(\eta_i)^2$ to $\pi(\eta)^2$. By the boundedness of $\{\pi(\eta_i)\}$, $\pi(\eta_i)$ converges strongly to $\pi(\eta)$. For each $\zeta \in \mathfrak{A}$, we have

$$\lim_i \|\pi'(\zeta)(\eta_i - \eta)\| = \lim_i \|(\pi(\eta_i) - \pi(\eta))\zeta\| = 0 .$$

If ζ is in $\mathfrak{A}$, then $F_m'\zeta$ is in $\mathfrak{D}(\Delta^{-\frac{1}{2}}) \cap \mathfrak{D}(\Delta^{\frac{1}{2}})$ and $\pi'(F_m'\zeta) = F_m'\pi'(\zeta)$; hence $F_m'\zeta$ is in $\mathfrak{A}' \cap \mathfrak{D}(\Delta^{\frac{1}{2}})$. By Theorem 9.1, we have, under suitable interpretation concerning domains,

$$\begin{aligned}\pi'(\Delta^{\frac{1}{2}}F_m'\zeta) &= \Delta^{\frac{1}{2}}\pi'(F_m'\zeta)\Delta^{-\frac{1}{2}} \\ &= \Delta^{\frac{1}{2}}F_m'\pi'(\zeta)\Delta^{-\frac{1}{2}}\end{aligned}$$

hence

$$\begin{aligned}\pi'(\Delta^{\frac{1}{2}}F_m'\zeta)F_kF_\ell' &= \exp(\tfrac{1}{2}H)\exp(\tfrac{1}{2}H')F_m'\pi'(\zeta)\exp(-\tfrac{1}{2}H')F'_\ell\exp(-\tfrac{1}{2}H)F_k \\ &= \exp(\tfrac{1}{2}H')F_m'\pi'(\zeta)\exp(-\tfrac{1}{2}H')F_kF_\ell' .\end{aligned}$$

Since $\pi'(\Delta^{\frac{1}{2}}F_m'\zeta)$ is bounded, we have

$$\pi'(\Delta^{\frac{1}{2}}F_m'\zeta)F_\ell' = \exp(\tfrac{1}{2}H')F_m'\pi'(\zeta)\exp(-\tfrac{1}{2}H')F_\ell' .$$

Therefore we have for $\zeta, \zeta' \in \mathfrak{A}$

$$\begin{aligned}(\exp(\tfrac{1}{2}H')F_m'\eta_i|\pi'(\zeta)\exp(-\tfrac{1}{2}H')F_\ell'\zeta') &= (\eta_i|\exp(\tfrac{1}{2}H')F_m'\pi'(\zeta)\exp(-\tfrac{1}{2}H')F_\ell'\zeta') \\ &= (\eta_i|\pi'(\Delta^{\frac{1}{2}}F_m'\zeta)F_\ell'\zeta') \\ &= (\pi'(\Delta^{\frac{1}{2}}F_m'\zeta)^*\eta_i|F_\ell'\zeta') \\ &= (\pi'((\Delta^{\frac{1}{2}}F_m'\zeta)^\flat)\eta_i|F_\ell'\zeta') ;\end{aligned}$$

hence we have

$$(\exp(\tfrac{1}{2}H')F'_m\eta \,|\, \pi'(\zeta)\exp(-\tfrac{1}{2}H')F'_\ell\zeta')$$

$$= \lim_i \, (\exp(\tfrac{1}{2}H')F'_m\eta_i \,|\, \pi'(\zeta)\exp(-\tfrac{1}{2}H')F'_\ell\zeta') \ .$$

Since $\{\pi'(\zeta)\exp(-\tfrac{1}{2}H')F'_\ell\zeta' \ ; \ \zeta,\zeta' \in \mathfrak{A}, \ \ell = 1,2,\dots\}$ is total in $\mathfrak{H}$ and $\|\exp(\tfrac{1}{2}H')\,F'_m\eta_i\|^2 \leq \|\exp(\tfrac{1}{2}H')\,\eta_i\|^2 = \tau(F_n x_i F_n)$ are bounded, the sequence $\{\exp(\tfrac{1}{2}H')F'_m\eta_i\}_{i\in I}$ converges weakly to $\exp(\tfrac{1}{2}H')\,F'_m\eta$. Therefore we get

$$\liminf \|\exp(\tfrac{1}{2}H')\eta_i\|^2 \geq \liminf \|\exp(\tfrac{1}{2}H')F'_m\eta_i\|^2$$

$$\geq \|\exp(\tfrac{1}{2}H')F'_m\eta\|^2$$

$$\geq \|\exp(\tfrac{1}{2}H')\eta\|^2 - \varepsilon/2$$

for sufficiently large m. Thus we can choose an $i \in I$ with $\|\exp(\tfrac{1}{2}H')\,\eta_i\|^2 \geq \|\exp(\tfrac{1}{2}H')\eta\|^2 - \varepsilon/2$, which implies the normality of the trace τ. Thus Theorem 14.1 has been proved.

Now, we shall consider a necessary condition for semi-finiteness. For the purpose, we employ the technique developed by Dixmier [4] and Pukanszky [26]. Let $\mathfrak{A}$ be an achieved generalized Hilbert algebra. Let $\mathfrak{A}_0$ be the modular Hilbert algebra constructed in §10. Suppose the left von Neumann algebra $\mathfrak{L}(\mathfrak{A})$, say M, is semi-finite. Let τ be a semi-finite faithful normal trace of M with definition ideal $\mathfrak{m}_\tau$. Then $\mathfrak{m}_\tau^{\frac{1}{2}} = \{x \in M : x^*x \in \mathfrak{m}_\tau\}$ is a strongly dense ideal of M and turns out to be an achieved Hilbert algebra whose left von Neumann algebra is spatially isomorphic to M. Let $\mathfrak{D}$ denote the set of all left bounded elements ξ with $\pi(\xi) \in \mathfrak{m}_\tau^{\frac{1}{2}}$.

Let $\mathfrak{m}$ denote the set of all $\pi(\xi)$ with $\xi \in \mathfrak{M}$. Then as in Lemma 14.1, $\mathfrak{m}$ is a left ideal of M. Let $\mathfrak{B}$ denote $\mathfrak{A} \cap \mathfrak{M} = \mathfrak{D}^{\#} \cap \mathfrak{M}$.

LEMMA 14.6. For $\xi \in \mathfrak{H}$ to be left bounded, it is necessary and sufficient that

$$\sup\{\|\pi'(\eta)\xi\|;\ \eta \in \mathfrak{A}_0,\ \|\eta\| \leq 1\} < +\infty .$$

<u>Proof</u>. The necessity is trivial. Suppose $\xi \in \mathfrak{H}$ satisfies the condition. Then we can define a bounded operator $\pi_0(\xi)$ on $\mathfrak{H}$ by $\pi_0(\xi)\eta = \pi'(\eta)\xi$, $\eta \in \mathfrak{A}_0$. If η is in $\mathfrak{A}'$, then there exists a sequence $\{\eta_n\}$ in $\mathfrak{A}_0$ with $\eta = \lim \eta_n$ and $\eta^{\flat} = \lim \eta_n^{\flat}$. For every $\zeta \in \mathfrak{A}_0$, we have

$$\begin{aligned}(\pi_0(\xi)\eta\,|\,\zeta) &= \lim(\pi_0(\xi)\eta_n\,|\,\zeta) = \lim(\pi'(\eta_n)\xi\,|\,\zeta)\\ &= \lim(\xi\,|\,\pi'(\eta_n^{\flat})\zeta) = \lim(\xi\,|\,\pi(\zeta)\eta_n^{\flat})\\ &= (\xi\,|\,\pi(\zeta)\eta^{\flat}) = (\xi\,|\,\pi'(\eta^{\flat})\zeta)\\ &= (\pi'(\eta)\xi\,|\,\zeta)\ ;\end{aligned}$$

hence $\pi'(\eta)\xi = \pi_0(\xi)\eta$ for every $\eta \in \mathfrak{A}'$. Therefore ξ satisfies the condition in Definition 14.1, that is, ξ is left bounded.

LEMMA 14.7. $\mathfrak{B}$ is dense in the Hilbert space $\mathfrak{D}^{\#}$. Hence, $\mathfrak{M}$ is dense in $\mathfrak{H}$.

<u>Proof</u>. Take a net $\{e_\alpha\}$ of self-adjoint elements in $\mathfrak{m}_\tau^{\frac{1}{2}}$ converging strongly to the identity. Let $\xi = \Sigma_{i=1}^n \eta_i^{\#}\zeta_i$ be an arbitrary element in $\mathfrak{A}^2$. Then by the dual assertion of Lemma 3.5,

$\mathfrak{A}$ contains the elements $\xi_\alpha = \Sigma_{i=1}^n \pi(\eta_i)^* e_\alpha \zeta_i$. Since $\mathfrak{m}_\tau^{\frac{1}{2}}$ is an ideal, $\pi(\xi_\alpha) = \Sigma_{i=1}^n \pi(\eta_i)^* e_\alpha \zeta_i$ belongs to $\mathfrak{m}_\tau^{\frac{1}{2}}$, so that ξ_α belongs to $\mathfrak{B}$. Since $\{e_\alpha\}$ converges to the identity,

$$\xi = \lim_\alpha \sum_{i=1}^n \pi(\eta_i)^* e_\alpha \zeta_i = \lim \xi_\alpha \ ;$$

$$\xi^\# = \sum_{i=1}^n \zeta_i^\# \eta_i = \lim_\alpha \sum_{i=1}^n \pi(\zeta_i)^* e_\alpha \eta_i$$

$$= \lim \xi_\alpha^\# \ .$$

Hence $\mathfrak{A}^2$ is contained in the closure of $\mathfrak{B}$ in the space $\mathfrak{D}^\#$. By Lemma 3.4, $\mathfrak{A}^2$ is dense in $\mathfrak{D}^\#$; hence so is $\mathfrak{B}$ in $\mathfrak{D}^\#$.

LEMMA 14.8. $\pi(\mathfrak{B})$ is dense in $L^2(M,\tau)$, where $L^2(M,\tau)$ is the Hilbert space constructed via τ as the completion of $\mathfrak{m}_\tau^{\frac{1}{2}}$. Therefore, $\mathfrak{m}$ is dense in $L^2(M,\tau)$.

The norm in $L^2(M,\tau)$ will be denoted by $\|\cdot\|_2$, that is, $\|x\|_2 = \tau(x^*x)^{\frac{1}{2}}$, $x \in \mathfrak{m}_\tau^{\frac{1}{2}}$.

Proof. If x is in $\mathfrak{m}_\tau^{\frac{1}{2}}$ and a, b are in $\pi(\mathfrak{A})$, then axb is in $\pi(\mathfrak{B})$ as in the proof of Lemma 14.7. If we choose a bounded net $\{e_\alpha\}$ of self-adjoint elements in $\pi(\mathfrak{A})$ converging strongly to the identity, then $e_\alpha x e_\alpha$ converges to x in $L^2(M,\tau)$ for every $x \in \mathfrak{m}_\tau^{\frac{1}{2}}$ because of the inequality:

$$\|x - e_\alpha x e_\alpha\|_2 \leq \|x - e_\alpha x\|_2 + \|e_\alpha x - e_\alpha x e_\alpha\|_2$$

$$\leq \|x - e_\alpha x\|_2 + \|e_\alpha\| \|x - x e_\alpha\|_2$$

$$= \|x - e_\alpha x\|_2 + \|e_\alpha\| \|x^* - e_\alpha x^*\|_2 .$$

This completes the proof.

LEMMA 14.9. π is a preclosed operator of $\mathfrak{M}$ onto $\mathfrak{m}$ with respect to the norm in $\mathfrak{M}$ and the L^2-norm in $\mathfrak{m}_\tau^{\frac{1}{2}}$, whose closure Π is non-singular.

Proof. Recall that τ can be written in the form:

$$\tau(x) = \sum_{i \in I} (x X_i | X_i), \quad X_i \in \mathfrak{H}, \quad x \in \mathfrak{m}_\tau .$$

Suppose that $\{\xi_n\}$ is a sequence in $\mathfrak{M}$ such that $\lim_n \xi_n = 0$ and $\{\pi(\xi_n)\}$ is a Cauchy sequence in $\mathfrak{m}_\tau^{\frac{1}{2}}$. Since the involution in $\mathfrak{m}_\tau^{\frac{1}{2}}$ is isometric, $\{\pi(\xi_n)^*\}$ is also a Cauchy sequence in $\mathfrak{m}_\tau^{\frac{1}{2}}$. Let a be the limit of $\{\pi(\xi_n)^*\}$ in $L^2(M,\tau)$. Recalling that $L^2(M,\tau)$ is imbedded isometrically in the Hilbert space $\Sigma_{i \in I} \oplus [MX_i]$ under the extended isometry of the map: $x \in \mathfrak{m}_\tau^{\frac{1}{2}} \mapsto \Sigma_{i \in I} \oplus x X_i$, let a_i denote the $[MX_i]$-component of a. Then we have $a_i = \lim_{n \to \infty} \pi(\xi_n)^* X_i$. For every $\eta \in \mathfrak{A}'$, we have

$$(a_i | \eta) = \lim_{n \to \infty} (\pi(\xi_n)^* X_i | \eta) = \lim_{n \to \infty} (X_i | \pi(\xi_n) \eta)$$

$$= \lim_n (X_i | \pi'(\eta) \xi_n) = 0 ,$$

so that a_i must be zero; hence $a = 0$. Therefore π is preclosed.

Next, suppose $\{\xi_n\}$ is a sequence in $\mathfrak{M}$ such that $\xi = \lim \xi_n$ and $\lim_n \pi(\xi_n) = 0$ in $\mathfrak{m}_\tau^{\frac{1}{2}}$. Then we also have $\lim_{n\to\infty} \pi(\xi_n)^* x_i = 0$. For every $\eta \in \mathfrak{A}'$, we have

$$(x_i | \pi'(\eta)\xi) = \lim (x_i | \pi'(\eta)\xi_n) = \lim (x_i | \pi(\xi_n)\eta)$$

$$= \lim (\pi(\xi_n)^* x_i | \eta) = 0 \ ,$$

so that we have $(\pi'(\mathfrak{A}')x_i | \xi) = 0$. By Theorem 3.1, $(M'x_i | \xi) = 0$. Since $\{x_i\}_{i\in I}$ is a separating family for M, it is a generating family for M'; hence $\xi = 0$. This completes the proof.

Now, let

$$\Pi = \Lambda K', \quad K' = (\Pi^*\Pi)^{\frac{1}{2}}$$

be the polar decomposition of Π. Then Λ is a unitary operator of $\mathfrak{H}$ onto $L^2(M,\tau)$, so that we have

(14.13) $$(K'\xi | K'\eta) = \tau(\pi(\xi)\pi(\eta)^*), \quad \xi,\eta \in \mathfrak{M} .$$

LEMMA 14.10. K' is affiliated with M'.

<u>Proof</u>. As seen above, $x\mathfrak{M} \subset \mathfrak{M}$ for every $x \in M$. Take and fix an element ξ in $\mathfrak{D}(K')$. Noticing that $\mathfrak{D}(K') = \mathfrak{D}(\Pi)$, we find a sequence $\{\xi_n\}$ in $\mathfrak{M}$ with $\xi = \lim_n \xi_n$ and $K'\xi = \lim_n K'\xi_n$. For each $x \in M$, $x\xi = \lim x\xi_n$ and

$$\|K'x\xi_n - K'x\xi_m\|^2 = \|x\pi(\xi_n - \xi_m)\|_2^2$$
$$\leq \|x\|^2 \|\pi(\xi_n) - \pi(\xi_m)\|_2^2$$
$$= \|x\|^2 \|K'\xi_n - K'\xi_m\|^2 \ ;$$

hence $\{K'x\xi_n\}$ converges. By closedness of K', $x\xi$ belongs to $\mathfrak{D}(K')$ and

$$K'x\xi = \lim_{n\to\infty} K'x\xi_n \ .$$

Hence $\mathfrak{D}(K')$ is invariant under the action of M.

For a unitary operator $u \in M$ and $\xi,\eta \in \mathfrak{M}$,

$$\begin{aligned}(K'u\xi|K'u\eta) &= \tau(\pi(u\xi)\pi(u\eta)^*) \\ &= \tau(u\pi(\xi)\pi(\eta)^*u^*) \\ &= \tau(\pi(\xi)\pi(\eta)^*) \\ &= (K'\xi|K'\eta) \ .\end{aligned}$$

Hence, for every pair ξ,η in $\mathfrak{D}(K')$ and a unitary operator $u \in M$,

$$(K'u\xi|K'u\eta) = (K'\xi|K'\eta) \ ;$$

hence $u^*K'^2u = K'^2$. Therefore $u^*K'u = K'$. This completes the proof.

LEMMA 14.11. $\mathfrak{M}$ is dense in the Hilbert space $\mathfrak{D}(K')$.

Proof. First we claim that if ξ is left bounded and η is in $\mathfrak{A}$ then $\pi(\eta)\xi$ is in $\mathfrak{A}$. In fact, $\pi(\eta)\xi$ is left bounded and for each $\zeta \in \mathfrak{A}'$,

$$\begin{aligned}(\pi(\eta)\xi|\zeta) &= (\xi|\pi(\eta^\#)\zeta) = (\xi|\pi'(\zeta)\eta^\#) \\ &= (\pi'(\zeta^\flat)\xi|\eta^\#) = (\pi(\xi)\zeta^\flat|\eta^\#) \\ &= (\zeta^\flat|\pi(\xi)^*\eta^\#) \ ;\end{aligned}$$

hence $\pi(\eta)\xi$ belongs to $\mathfrak{D}^{\#}$ and

$$(\pi(\eta)\xi)^{\#} = \pi(\xi)^{*}\eta^{\#} .$$

Now, take an element ξ in $\mathfrak{M}$ and choose a net $\{\eta_\alpha\}$ in $\mathfrak{A}$ such that $\|\pi(\eta_\alpha)\| \leq 1$ and $\{\pi(\eta_\alpha)\}$ converges strongly to the identity. Then $\pi(\eta_\alpha)\xi$ is in $\mathfrak{B}$ and

$$\xi = \lim_{\alpha} \pi(\eta_\alpha)\xi ;$$

$$\|K'\xi - K'\pi(\eta_\alpha)\xi\|^2 = \|\pi(\xi) - \pi(\eta_\alpha)\pi(\xi)\|_2^2 ;$$

hence

$$K'\xi = \lim_{\alpha} K'\pi(\eta_\alpha)\xi .$$

This completes the proof.

Define a self-adjoint positive operator K by:

$$K = JK'J .$$

Then by Theorem 10.1 K is affiliated with M. By (14.13) we have, for $\xi,\eta \in \mathfrak{B}$,

$$\begin{aligned}(K'\xi | K'\eta) &= (K'\eta^{\#} | K'\xi^{\#}) = (K'J\Delta^{\frac{1}{2}}\eta | K'J\Delta^{\frac{1}{2}}\xi) \\ &= (JK'J\Delta^{\frac{1}{2}}\xi | JK'J\Delta^{\frac{1}{2}}\eta) \\ &= (K\Delta^{\frac{1}{2}}\xi | K\Delta^{\frac{1}{2}}\eta) ;\end{aligned}$$

therefore

(14.14) $$(K\Delta^{\frac{1}{2}}\xi | K\Delta^{\frac{1}{2}}\eta) = (K'\xi | K'\eta), \quad \xi,\eta \in \mathfrak{M} .$$

Let ρ be the representation of M on $L^2(M,\tau)$ defined by:

$$\rho(x)y = xy, \quad x \in M, \quad y \in \mathfrak{m}_\tau^{\frac{1}{2}} .$$

Let ρ' be the anti-representation of M on $L^2(M,\tau)$ defined by:

$$\rho'(x)y = yx, \quad x \in M, \quad y \in \mathfrak{m}_\tau^{\frac{1}{2}} .$$

Then we have the commutation relation:

$$\rho(M)' = \rho'(M), \quad \rho'(M)' = \rho(M) .$$

LEMMA 14.12. The unitary operator Λ induces a spatial isomorphism between M and $\rho(M)$ such that

$$\Lambda\pi(\xi)\Lambda^{-1} = \rho \cdot \pi(\xi), \quad \xi \in \mathfrak{M} .$$

<u>Proof</u>. For $\xi,\eta \in \mathfrak{M}$, we have

$$\pi(\xi\eta) = \pi(\xi)\pi(\eta) = \rho \cdot \pi(\xi)\Pi\eta;$$

hence

$$\Lambda\pi(\xi)K'\eta = \Lambda K'\pi(\xi)\eta = \Pi(\xi\eta)$$

$$= \rho \cdot \pi(\xi)\Pi\eta$$

$$= \rho \cdot \pi(\xi)\Lambda K'\eta .$$

Since $K'\mathfrak{M}$ is dense in $\mathcal{H}$ and $\pi(\mathfrak{M})$ is also dense in $L^2(M,\tau)$, we have

$$\Lambda\pi(\xi) = \rho \cdot \pi(\xi)\Lambda, \quad \xi \in \mathfrak{M} .$$

This completes the proof.

LEMMA 14.13. If $\Pi\xi$, $\xi \in \mathfrak{D}(K')$, is bounded, that is, $\Pi\xi \in \mathfrak{m}_\tau^{\frac{1}{2}}$, then ξ is left bounded; hence $\xi \in \mathfrak{M}$.

Proof. Since $\mathfrak{B}$ is dense in the Hilbert space $\mathfrak{D}(K') = \mathfrak{D}(\Pi)$, there is a sequence $\{\xi_n\}$ in $\mathfrak{B}$ with

$$\xi = \lim \xi_n \quad \text{and} \quad \Pi\xi = \lim_n \pi(\xi_n) .$$

Put $x = \Pi\xi$. For each $\eta \in \mathfrak{A}_0$, we have

$$\pi'(\eta)\xi = \lim_n \pi'(\eta)\xi_n ;$$

$$\lim \|\pi(x\eta) - \pi(\pi'(\eta)\xi_n)\|_2 = \lim \|x\pi(\eta) - \pi(\xi_n)\pi(\eta)\|_2$$

$$= 0 ;$$

hence, by closedness of Π,

$$\Pi\pi'(\eta)\xi = \pi(x\eta) .$$

Since Π is non-singular, we have

$$\pi'(\eta)\xi = x\eta ;$$

hence ξ is left bounded and $\pi(\xi) = x$.

For a bounded Borel function f defined on the positive real numbers, we define an operator $H_f' \in M'$ by $H_f' = f(K')$. Then there is an operator $H_f \in M$ with

$$\rho'(H_f) = \Lambda H_f' \Lambda^{-1} . \tag{14.15}$$

LEMMA 14.14. $H_f'\mathfrak{M} \subset \mathfrak{M}$ and

$$\pi(H_f'\xi) = \pi(\xi)H_f, \quad \xi \in \mathfrak{M} .$$

<u>Proof</u>. By the definition of H_f', $H_f'\mathfrak{D}(K') \subset \mathfrak{D}(K')$; hence if ξ is in $\mathfrak{M}$, then $H_f'\xi$ is in $\mathfrak{D}(K')$. Furthermore, we have

$$\begin{aligned} \Pi H_f'\xi &= \Lambda K'H_f'\xi = \Lambda H_f'K'\xi \\ &= \Lambda H_f'\Lambda^{-1}\Lambda K'\xi = \rho'(H_f)\pi(\xi) \\ &= \pi(\xi)H_f . \end{aligned}$$

Therefore Lemma 14.13 yields our assertion.

LEMMA 14.15. $H_f\mathfrak{A} \subset \mathfrak{A}$ and

$$\pi(H_f'\xi) = \pi(\xi)H_f, \quad \xi \in \mathfrak{A} .$$

<u>Proof</u>. Take and fix an element $\xi \in \mathfrak{A}$. For every $x \in \mathfrak{m}_\tau^{\frac{1}{2}}$, $x\xi$ belongs to $\mathfrak{M}$. By Lemma 14.14,

$$\pi(H_f'x\xi) = \pi(x\xi)H_f = x\pi(\xi)H_f .$$

Choosing a net $\{x_\alpha\}$ in $\mathfrak{m}_\tau^{\frac{1}{2}}$ with $\lim x_\alpha = 1$, we have, for each $\eta \in \mathfrak{A}'$,

$$\begin{aligned} \pi'(\eta)H_f'\xi &= \lim_\alpha \pi'(\eta)H_f'x_\alpha\xi \\ &= \lim_\alpha x_\alpha\pi(\xi)H_f\eta \\ &= \pi(\xi)H_f\eta ; \end{aligned}$$

hence $H_f'\xi$ is left bounded and $\pi(H_f'\xi) = \pi(\xi)H_f$. For each $\eta_1,\eta_2 \in \mathfrak{A}'$, we have

$$(H_f'\xi|\eta_1\eta_2^\flat) = (\pi'(\eta_2)H_f'\xi|\eta_1)$$

$$= (\pi(\xi)H_f\eta_2|\eta_1) = (\eta_2|H_f^*\pi(\xi^\#)\eta_1)$$

$$= (\eta_2\eta_1^\flat|H_f^*\xi^\#) ;$$

hence by Lemma 3.3, $H_f'\xi$ is in $\mathfrak{D}^\#$ and

(14.16) $$(H_f'\xi)^\# = H_f^*\xi^\# .$$

This completes the proof.

From (14.16) it follows that

$$\Delta^{\frac{1}{2}}H_f'\xi = JH_f^*J\Delta^{\frac{1}{2}}\xi, \quad \xi \in \mathfrak{A} .$$

If ξ is in $\mathfrak{D}^\#$, then there exists a sequence $\{\xi_n\}$ in $\mathfrak{A}$ with $\xi = \lim \xi_n$ and $\Delta^{\frac{1}{2}}\xi = \lim \Delta^{\frac{1}{2}}\xi_n$, so that

$$H_f'\xi = \lim_n H_f'\xi_n ;$$

$$\lim \Delta^{\frac{1}{2}}H_f'\xi_n = \lim JH_f^*J\Delta^{\frac{1}{2}}\xi_n$$

$$= JH_f^*J\Delta^{\frac{1}{2}}\xi ;$$

hence $H_f'\xi$ is in $\mathfrak{D}^\#$ and

$$\Delta^{\frac{1}{2}}H_f'\xi = JH_f^*J\Delta^{\frac{1}{2}}\xi, \quad \xi \in \mathfrak{D}^\# .$$

Hence $H_f'\mathfrak{D}^\# \subset \mathfrak{D}^\#$. Since f is arbitrary, we may choose f as $f(\lambda) = (\lambda + i)(\lambda - i)^{-1}$. Then H_f' is the Cayley transform of K' and $H_f'^{-1} = H_{\bar{f}}$. Therefore, in this case we have

$$\Delta^{\frac{1}{2}}H_f' = JH_f^*J\Delta^{\frac{1}{2}} ;$$

hence

$$\Delta^{\frac{1}{2}} = JH_f^* J\Delta^{\frac{1}{2}} H_f'^* .$$

Considering adjoint operator, we have

$$\Delta^{\frac{1}{2}} = H_f' \Delta^{\frac{1}{2}} JH_f J ,$$

so that

$$\Delta = \Delta^{\frac{1}{2}}\Delta^{\frac{1}{2}} = H_f' \Delta H_f'^* .$$

Therefore, Δ and H_f' commute with each other, and then Δ commutes with K'. Hence Δ commutes also with K. Therefore, $K\Delta^{\frac{1}{2}}$ has a unique self-adjoint extension, say K_1'. By (14.14), we have

$$(K_1'\xi | K_1'\eta) = (K'\xi | K'\eta), \quad \xi, \eta \in \mathfrak{M} .$$

If ξ is in $\mathfrak{D}(K')$, then there is a sequence $\{\xi_n\}$ in $\mathfrak{M}$ with $\xi = \lim \xi_n$ and $K'\xi = \lim K'\xi_n$. Then

$$\lim_{n,m\to\infty} \|K_1'\xi_n - K_1'\eta_m\|^2 = \lim_{n,m\to\infty} \|K'\xi_n - K'\xi_m\|^2 = 0 ,$$

so that $\{K_1'\xi_n\}$ converges to $K_1'\xi$ by the closedness of K_1'. Therefore, $\mathfrak{D}(K_1')$ contains $\mathfrak{D}(K')$ as a closed subspace. If $\mathfrak{D}(K')$ is a proper subspace of $\mathfrak{D}(K_1')$, then there exists a non-zero vector $\xi \in \mathfrak{D}(K_1')$ with $\langle \xi | \eta \rangle_{K_1'} = 0$ for every $\eta \in \mathfrak{D}(K')$. Then we have

$$\begin{aligned} 0 &= (\xi|\eta) + (K_1'\xi | K_1'\eta) \\ &= (\xi|\eta) + (K_1'\xi | K'\eta) \end{aligned}$$

for every $\eta \in \mathcal{D}(K')$; hence

$$(K_1'\xi|K'\eta) = -(\xi|\eta), \quad \eta \in \mathcal{D}(K'),$$

which means that $K_1'\xi$ belongs to $\mathcal{D}(K')$ and

$$K'K_1'\xi = -\xi .$$

But this is impossible because K' and K_1' are commutative and positive. Therefore $\mathcal{D}(K_1')$ coincides with $\mathcal{D}(K')$ and

$$(K'\xi|K'\eta) = (K'\xi|K'\eta), \quad \xi,\eta \in \mathcal{D}(K') .$$

Therefore $K_1'^2 = K'^2$ and then their positivity implies $K_1' = K'$. Therefore we have

$$(14.17) \qquad \Delta^{\frac{1}{2}} = K^{-1} \cdot K' ,$$

where $K^{-1} \cdot K'$ means the closure of $K^{-1}K'$. Thus the associated modular automorphism σ_t of M is induced by the strongly continuous one parameter unitary group K^{-2it}, $-\infty < t < +\infty$, in M. After all, we get the following:

THEOREM 14.2. If the left von Neumann algebra of a modular (or generalized) Hilbert algebra is semi-finite, then the associated modular automorphism group σ_t is inner.

Combining Theorem 14.1, 14.2 and (14.17), we get the following.

COROLLARY 14.1. If the modular automorphism group σ_t is inner, then the one-parameter group Δ^{it}, $-\infty < t < +\infty$, is represented into the form:

(14.18) $$\Delta^{it} = \Gamma(t)\Gamma'(t), \quad J\Gamma(t)J = \Gamma'(-t) ,$$

where $\Gamma(t)$ (resp. $\Gamma'(t)$) is a one-parameter unitary group in the left von Neumann algebra (resp. the commutant of the left von Neumann algebra).

COROLLARY 14.2. Let φ be a faithful normal positive linear functional of a von Neumann algebra M. In order that M is semi-finite, it is necessary and sufficient that the modular automorphism σ_t associated with φ is inner.

Now we shall study the modular automorphism σ_t of a semi-finite von Neumann algebra M associated with a faithful normal positive linear functional φ. Let τ be a faithful normal semi-finite trace of M with definition ideal $\mathfrak{m}_\tau$. Then there exists a unique positive self-adjoint operator h in $L^1(M,\tau)$ such that

(14.19) $$\varphi(x) = \tau(xh), \quad x \in M .$$

Put $u(t) = h^{it}$, $-\infty < t < \infty$. Then $u(t)$ is a one-parameter unitary group in M. Define a one-parameter automorphism group σ_t of M by:

(14.20) $$\sigma_t(x) = u(t)xu(-t), \quad x \in M .$$

Since $u(t)hu(-t) = h$, φ is invariant under σ_t.

Now we claim that for each $x \in M$, the function:

$$z \in C \mapsto h^{iz+1}xh^{-iz} \in L^1(M,\tau)$$

is defined and analytic in the strip $0 < \mathrm{Im}\, z < 1$ and is continuous on the strip $0 \leq \mathrm{Im}\, z \leq 1$. Put $z = s + it$, $s,t \in R$ and $0 \leq t \leq 1$. Then we have

$$h^{iz+1}xh^{-iz} = h^{1-t}h^{is}xh^{-is}h^{t} .$$

Since $h^{1-t} \in L^{1/(1-t)}(M,\tau)$ and $h^{t} \in L^{1/t}(M,\tau)$ and

$$\|h^{1-t}\|_{1/(1-t)} = \|h\|_1^{1-t} \quad \text{and} \quad \|h^t\|_{1/t} = \|h\|_1^t ,$$

where $\|\cdot\|_p$ means the norm in $L^p(M,\tau)$, we have

$$\|h^{1-t}h^{is}xh^{-is}h^{t}\|_1 \leq \|x\|\|h\|_1 .$$

Let $h = \int_0^\infty \lambda de(\lambda)$ be the spectral decomposition of h. Put $e_n = \int_{1/n}^n de(\lambda)$ for each $n = 1,2,\ldots$. Then $\{e_n\}$ converges strongly to the identity. Since for each $x \in M$, we have

$$\lim \|h^{iz+1}e_nxe_nh^{-iz} - h^{iz+1}xh^{-iz}\|_1 = 0 ;$$

$$\|h^{iz+1}e_nxe_nh^{-iz}\| \leq \|e_nxe_n\|\|h\|_1$$

for each $z \in C$ with $0 < \mathrm{Im}\, z < 1$, in order to prove the analyticity of the function $z \in C \rightarrow h^{iz+1}xh^{-iz}$, it is sufficient to show that the function: $z \rightarrow h^{iz+1}e_nxe_nh^{-iz}$ is analytic. But he_n is bounded and invertible in e_nMe_n, so that the function: $z \rightarrow (he_n)^{iz}$ is uniformly analytic on the whole plane C as an e_nMe_n-valued function; hence $h^{iz+1}e_nxe_nh^{-iz}$ is an analytic e_nMe_n-valued function of z on the whole plane. Since the injection of e_nMe_n into $L^1(M,\tau)$ is continuous, our assertion follows.

Now we can conclude that φ satisfies the KMS-boundary condition for σ_t as follows. Take an arbitrary pair x, y in M. Then the function: $z \in C \mapsto yh^{iz+1}xh^{-iz}$ is an $L^1(M,\tau)$-valued analytic function in the strip $0 < \mathrm{Im}\, z < 1$; hence so is the function $f(z) = \tau(yh^{iz+1}xh^{-iz})$. For each $t \in R$, if we put

$$\varphi(\sigma_t(x)y) = \tau(h^{it}xh^{-it}yh) = \tau(yh^{it+1}xh^{-it}) ;$$

$$\varphi(y\sigma_t(x)) = \tau(yh^{it}xh^{-it+1}) ,$$

then we have $f(t) = \varphi(\sigma_t(x)y)$ and $f(t + i) = \varphi(y\sigma_t(x))$. Thus φ satisfies the KMS-boundary condition for the one parameter automorphism group σ_t. Therefore, by Theorem 13.2, σ_t is the modular automorphism group of M associated with φ.

COROLLARY 14.3. Every type III factor with separable predual has an outer automorphism.

<u>Proof</u>. Let M be a type III factor with separable predual M_*. Take a faithful normal state φ of M, where the existence of such φ follows from the separability of M_*. Let σ_t be the modular automorphism group associated with φ. If every σ_t is inner, then by [24; Theorem 4.13] there exists a strongly continuous one parameter unitary group $\Gamma(t)$ in M such that $\sigma_t(x) = \Gamma(t)x\Gamma(t)^{-1}$ for every $x \in M$. Then by Theorem 14.1, M must be semi-finite, which is contradiction. Therefore, $\{\sigma_t\}$ contains an outer automorphism of M. This completes the proof.

15. The Radon-Nikodym Theorem and the Modular Automorphism Group

As an application of our theory, we shall study the non-commutative extension of the so called Radon-Nikodym Theorem in the measure theory

Let M be a von Neumann algebra and φ_0 a faithful normal positive linear functional of M. Let σ_t be the modular automorphism group of M associated with φ_0. Considering the cyclic representation of M induced by φ_0, we may assume that M acts on a Hilbert space $\mathfrak{H}$ with a separating and generating vector ξ_0 such that $\varphi_0(x) = (x\xi_0|\xi_0)$, $x \in M$. As in Theorem 12.1, $\mathfrak{A} = M\xi_0$ turns out to be an achieved generalized Hilbert algebra. In this case, we have $\mathfrak{A}' = M'\xi_0$. For each $\xi \in \mathfrak{H}$, define a normal linear functional φ'_ξ (resp. φ_ξ) on M' (resp. M) by:

$$\varphi'_\xi(x') = (x'\xi|\xi_0), \quad x' \in M ;$$

(15.1)

$$\varphi_\xi(x) = (x\xi|\xi_0), \quad x \in M .$$

Let V' (resp. V) be the set of all φ'_ξ (resp. φ_ξ), $\xi \in \mathfrak{H}$. Then we easily get the following:

LEMMA 15.1. The following (i) and (ii) (resp. (i') and (ii')) are equivalent:

(i) ξ belongs to $\mathfrak{D}^\#$, (i') ξ belongs to $\mathfrak{D}^\flat$;

(ii) φ'^*_ξ belongs to V', (ii') φ^*_ξ belongs to V.

If ξ is in $\mathfrak{D}^\#$ (resp. $\mathfrak{D}^\flat$), then we have

$$(15.2) \qquad \varphi'^*_\xi = \varphi'_{\xi^\#} \qquad (\text{resp. } \varphi^*_\xi = \varphi_{\xi^\flat}) .$$

Let $\mathcal{P}^\#$ (resp. $\mathcal{P}^\flat$) denote the set of all $\xi \in \mathfrak{D}^\#$ (resp. $\xi \in \mathfrak{D}^\flat$) such that $\varphi'_\xi \geq 0$ (resp. $\varphi_\xi \geq 0$). Then $\mathcal{P}^\#$ and $\mathcal{P}^\flat$ are both cones in $\mathfrak{H}$.

LEMMA 15.2. $\mathfrak{D}^\#$ and $\mathfrak{D}^\flat$ are both algebraically spanned by $\mathcal{P}^\#$ and $\mathcal{P}^\flat$ respectively. Furthermore, $\mathcal{P}^\#$ and $\mathcal{P}^\flat$ are dual cones of each other in the following sense:

(i) A vector $\xi \in \mathfrak{H}$ belongs to $\mathcal{P}^\#$ if and only if $(\xi|\eta) \geq 0$ for every $\eta \in \mathcal{P}^\flat$;

(ii) A vector $\eta \in \mathfrak{H}$ belongs to $\mathcal{P}$ if and only if $(\xi|\eta) \geq 0$ for every $\xi \in \mathcal{P}^\#$.

<u>Proof</u>. Because of symmetry, we have only to prove the assertion for $\mathfrak{D}^\#$ and $\mathcal{P}^\#$. Since $\mathfrak{D}^\#$ is spanned by its self-adjoint part, we shall prove that every self-adjoint $\xi \in \mathfrak{D}^\#$, i.e. $\xi = \xi^\#$, can be represented by the difference $\xi_1 - \xi_2$ of two elements $\xi_1, \xi_2 \in \mathcal{P}^\#$. Define two actions of each $a \in M$ (resp. M') on $\varphi \in M_*$ (resp. M'_*) by

$$(15.3) \quad \langle x, a\varphi\rangle = \langle xa, \varphi\rangle \quad \text{and} \quad \langle x, \varphi a\rangle = \langle ax, \varphi\rangle, \quad x \in M \quad (\text{resp. } x \in M'),$$

where $\langle x, \varphi\rangle$ means the value of φ at x. Then we have

$$a\varphi_\xi = \varphi_{a\xi}, \quad a'\varphi'_\xi = \varphi'_{a'\xi}$$

for each $a \in M$ (resp. $a' \in M'$). If φ'_ξ is self-adjoint, then there exists a projection $e' \in M'$ such that

$$e'\varphi'_\xi \geq 0 \quad \text{and} \quad (1 - e')\varphi'_\xi \leq 0 .$$

Hence putting $\xi_1 = e'\xi$ and $\xi_2 = -(1 - e')\xi$, we get the desired decomposition: $\xi = \xi_1 - \xi_2$, $\xi_1, \xi_2 \in P^\#$.

Take a $\xi \in H$. Suppose that $(\xi|\eta) \geq 0$ for each $\eta \in P^\flat$. Since $M'_+\xi_0 \subset P^\flat$, where M'_+ denotes the set of all positive elements in M', we have, for each $x \in M'$,

$$0 \leq (\xi|x^*x\xi_0) = (x^*x\xi|\xi_0) ,$$

so that $\varphi'_\xi \geq 0$; and then ξ belongs to $P^\#$.

Take arbitrarily a $\xi \in P^\#$ and an $\eta \in P^\flat$. We shall prove that $(\xi|\eta) \geq 0$. Define an operator h_0 on $\mathfrak{A}'$ by:

$$h_0 x\xi_0 = x\xi, \quad x \in M' .$$

Then h_0 is a densely defined positive operator commuting with every unitary operator in M'. Let h denote the Friedrich's self-adjoint extension of h_0 (see [6, Chapter XII]). Then h commutes with every unitary operator in M' because the relevant positive sesquilinear form, which is used to extend h_0, is invariant under the action of every unitary operator in M'. Hence h is affiliated with M. Let $h = \int_0^\infty \lambda de(\lambda)$ denote the spectral decomposition of h. Put $h_n = \int_0^n \lambda de(\lambda)$, $n = 1,2,\ldots$. Then $h_n\xi_0$ converges strongly to $h\xi_0 = \xi$. Thus we get, for each $\eta \in P^\flat$,

$$(\xi|\eta) = \lim (h_n\xi|\eta) = \lim (\xi_0|h_n\eta)$$

$$= \lim \overline{(h_n\eta|\xi_0)} \geq 0 ,$$

since $\varphi_\eta \geq 0$. This completes the proof.

For each pair ξ, η in H, define a normal linear functional

$\omega_{\xi,\eta}$ (resp. $\omega'_{\xi,\eta}$) of M (resp. M') by:

$$\langle x,\omega_{\xi,\eta}\rangle = (x\xi|\eta), \quad x \in M ;$$

(15.4)

$$\langle x,\omega'_{\xi,\eta}\rangle = (x\xi|\eta), \quad x \in M' .$$

Then clearly we have $\varphi_\xi = \omega_{\xi,\xi_0}$ and $\varphi'_\xi = \omega'_{\xi,\xi_0}$. Furthermore

$$\omega_{\xi,\eta}{}^* = \omega_{\eta,\xi} \text{ and } (\omega'_{\xi,\eta})^* = \omega'_{\eta,\xi} .$$

LEMMA 15.3. If $\omega_{\xi_1,\xi_1} = \omega_{\xi_2,\xi_2}$ for $\xi_1,\xi_2 \in P^\#$, then we have $\xi_1 = \xi_2$.

<u>Proof</u>. For each $x \in M$, $\|x\xi_1\| = \|x\xi_2\|$, so that there exists a partial isometry $u' \in M'$ such that

$$u'x\xi_1 = x\xi_2 \text{ and } u'[M\xi_1]^\perp = 0 .$$

In particular, $u'\xi_1 = \xi_2$ and $u'^*\xi_2 = \xi_1$. Hence $\varphi'_{\xi_2} = \varphi'_{u'\xi_1} = u'\varphi'_{\xi_1}$, and $\varphi'_{\xi_1} = u'^*\varphi'_{\xi_2}$. Since φ'_{ξ_1} and φ'_{ξ_2} are both positive, $\varphi'_{\xi_2} = \varphi'_{\xi_1}$, because of the unicity of the polar decomposition of normal linear functionals (see [3, Theorem 4, p. 61]). Hence we get $\xi_1 = \xi_2$.

LEMMA 15.4. If φ is a normal positive linear functional of M and $\varphi \leq \varphi_0$, then there exists a unique element $h \in M$ such that $\varphi = h\varphi_0 h$ and $0 \leq h \leq 1$.

The existence of such h was first proved by S. Sakai in [16]. We shall for the sake of completeness, give a proof which is different from his.

Proof. By assumption, there exists an $h' \in M'$ such that $\varphi(x) = (xh'\xi_0 | h'\xi_0)$ and $0 \leq h' \leq 1$. Put $\xi = h'\xi_0$. Let $\varphi'_\xi = u'\psi$ be the polar decomposition of φ'_ξ in M'_*. Then $\psi = u'^*\varphi'_\xi = \varphi'_{u'^*\xi}$. Put $\xi_1 = u'^*\xi$. Then ξ_1 is in $P^\#$ since $\psi = \varphi'_{\xi_1}$ is positive. Since $\varphi'_\xi = u'\varphi'_{\xi_1} = \varphi'_{u'\xi_1}$, we have $\xi = u'\xi_1$; hence we have, for each $x \in M$,

$$\varphi(x) = (x\xi|\xi) = (x\xi|u'\xi_1) = (xu'^*\xi|\xi_1) = (x\xi_1|\xi_1) .$$

Since $\varphi = \omega_{\xi_1,\xi_1}$ is majorized by φ_0, ξ_1 is in $\mathfrak{A}'$ and $\|\pi'(\xi_1)\| \leq 1$. Hence ξ_1 is in $\mathfrak{A}' \cap P^\#$. We claim that $\mathfrak{A}' \cap P^\# \subset \mathfrak{A}^\#$. In fact, noticing that $\xi = \xi^\# = J\Delta^{\frac{1}{2}}\xi$, we have

$$\xi^\flat = \Delta^{\frac{1}{2}}J\xi = \Delta^{\frac{1}{2}} \cdot \Delta^{\frac{1}{2}}\xi = \Delta\xi \in \mathfrak{A}' ,$$

so that ξ and $\Delta\xi$ are both in $\mathfrak{A}'$. Hence by the definition of $\mathfrak{A}^\#$, ξ is in $\mathfrak{A}^\#$. Furthermore, putting $\eta = \xi + \xi^\flat$ we have $\xi = (\Delta + 1)^{-1}\eta$; hence by Lemma 8.1

$$\|\pi(\xi)\| \leq \tfrac{1}{2}\|\pi'(\eta)\| \leq \tfrac{1}{2}(\|\pi'(\xi)\| + \|\pi'(\xi)^*\|) = 1 .$$

Putting $h = \pi(\xi)$, we get the desired h. The unicity of h follows from Lemma 15.3.

THEOREM 15.1. For every normal positive linear functional φ of M, there exists a unique $\xi \in P^\#$ such that $\varphi = \omega_{\xi,\xi}$. Hence there exists a self-ajoint positive operator h affiliated with

M such that $\varphi(x) = (xh\xi_0|h\xi_0)$ for every $x \in M$. If φ is majorized by φ_0, then the above h is unique and $0 \leq h \leq 1$.

This result is an improved version of the well-known theorem which is found, for example, in [3, Theorem 4, p. 222] and [23]. As seen in [23], we can claim, without making use of neither von Neumann's BT-Theorem nor its alternation in [3; Lemma 3, p. 220], that φ has the form $\varphi = \omega_{\xi,\xi}$ for some $\xi \in \mathfrak{H}$. Let $\varphi'_\xi = u'\psi$ be the polar decomposition of φ'_ξ in M'_*. Then $\psi = u'^*\varphi'_\xi = \varphi'_{u'^*\xi}$. Put $\xi_1 = u'^*\xi$. Then $\varphi'_{\xi_1} = \psi$ is positive, so that ξ_1 belongs to $P^\#$. By the equality:

$$\varphi'_\xi = u'\psi = u'\varphi'_{\xi_1} = \varphi'_{u'\xi_1},$$

we have $\xi = u'\xi_1$. Hence we have

$$\varphi(x) = (x\xi|\xi) = (x\xi|u'\xi_1) = (xu'^*\xi|\xi_1) = (x\xi_1|\xi_1)$$

for every $x \in M$.

Therefore, the ξ in Theorem 15.1 can be regarded as $(d\varphi/d\varphi_0)^{\frac{1}{2}}$ by the unicity. The h in Theorem 15.1 would presumably be unique. Unfortunately, the Friedrick's extension of a positive symmetric operator may not be a unique positive self-adjoint extension. To establish the unicity of h, we should study more carefully all possible extensions of the operator h_0 in the proof of Lemma 15.4.

Now we shall define the commutativity of two normal positive linear functionals.

DEFINITION 15.1. Two normal positive linear functionals φ and ψ on M are said to <u>commute</u> if the two normal linear functionals

$\varphi + i\psi$ and $\varphi - i\psi$ have the same absolute value.

DEFINITION 15.2. For each $\varphi \in M_*$, the smallest projection e (resp. f) in M with $e\varphi = \varphi$ (resp. $\varphi f = \varphi$) is called the left (resp. right) support projection of φ and denoted by $s_\ell(\varphi)$ (resp. $s_r(\varphi)$). Of course, if φ is self-adjoint, then $s_\ell(\varphi) = s_r(\varphi)$, which will therefore be denoted by $s(\varphi)$.

LEMMA 15.5. For each $\varphi \in M_*$, $s_\ell(\varphi)$ and $s_r(\varphi)$ are the projections such that

$$[\varphi M] = s_\ell(\varphi)M_* \quad \text{and} \quad [M\varphi] = M_* s_r(\varphi) ,$$

where $[\cdot]$ means the closure of $\cdot$ in M_*.

<u>Proof</u>. By symmetry, we have only to prove the assertion for $s_\ell(\varphi)$. Since $[\varphi M]$ is a right invariant subspace of M_*, $[\varphi M]$ has the form eM_* for some projection $e \in M$ by [19]. Since φ is in eM_*, $e\varphi = \varphi$. Conversely, if $f\varphi = \varphi$ for some projection $f \in M$, then φ is in fM_*; hence $[\varphi M]$ is contained in fM_*. Therefore $e \leq f$. Thus $e = s_\ell(\varphi)$.

LEMMA 15.6. For each $\varphi \in M_*$, let $\varphi = u|\varphi|$ be the polar decomposition of φ in M_*. Then we have

$$u^*u = s_r(\varphi) \quad \text{and} \quad uu^* = s_\ell(\varphi) . \tag{15.5}$$

Furthermore, we have

$$s_r(\varphi) = s(|\varphi|) \quad \text{and} \quad s_\ell(\varphi) = s(|\varphi^*|) .$$

<u>Proof</u>. By definition of the polar decomposition (see [3, pp. 61-62]), $u^*u = s(|\varphi|)$. By equalities $u^*\varphi = |\varphi|$ and $\varphi = u|\varphi|$, $\varphi e = \varphi$ if and only if $|\varphi|e = |\varphi|e$ for each projection $e \in M$, so that $s_r(\varphi) = s(|\varphi|)$. If $eu = u$ for some projection $e \in M$, then $e\varphi = \varphi$, so that $uu^* \geq s_\ell(\varphi)$. If $e\varphi = \varphi$ for some projection $e \in M$, then $\varphi = u_1|\varphi|$ for $u_1 = eu$. By the unicity of polar decomposition, we have $u_1 = u$; hence $e \geq s\,(\varphi)$. By the equalities $\varphi^* = |\varphi|u^* = u^*u|\varphi|u^*$ and $u\varphi^* = u|\varphi|u^*$, we have $|\varphi^*| = u|\varphi|u^*$. Hence $s(|\varphi^*|) = us(|\varphi|)u^* = uu^*$. This completes the proof.

LEMMA 15.7. For each $\psi \in M_*^+$, where M_*^+ denotes the positive part of M_*, we have

$$s_\ell(\varphi_0 + i\psi) = s_r(\varphi_0 + i\psi) = 1 ;$$

$$s_\ell(\varphi_0 - i\psi) = s_r(\varphi_0 - i\psi) = 1 .$$

<u>Proof</u>. Put $e = s_\ell(\varphi_0 + i\psi)$. By the equality $e(\varphi_0 + i\psi) = \varphi_0 + i\psi$, we have

$$\begin{aligned}\langle e, \varphi_0 + i\psi\rangle &= \langle 1, e(\varphi_0 + i\psi)\rangle \\ &= \langle 1, \varphi_0 + i\psi\rangle ,\end{aligned}$$

so that

$$\langle 1 - e, \varphi_0\rangle + i\langle 1 - e, \psi\rangle = 0 .$$

Since φ_0 is faithful, we have $e = 1$. Similarly, we can get the other assertions. This completes the proof.

LEMMA 15.8. For each $x \in M$, $\sigma_t(x) = x$, $-\infty < t < \infty$, if and only if $x\varphi_0 = \varphi_0 x$.

Proof. For each $y \in M$, consider a function $F(z)$ holomorphic in the strip $0 < \operatorname{Im} z < 1$ and with boundary values:

$$F(t) = \varphi_0(\sigma_t(x)y); \; F(t+i) = \varphi_0(y\sigma_t(x)) .$$

$F(t)$ is constant if and only if $F(z)$ is a constant function, which is equivalent, by Strum's Theorem, to saying that $F(t) = F(t+i)$ for all real t, that is, $\varphi_0(\sigma_t(x)y) = \varphi_0(y\sigma_t(x))$. Noticing that φ_0 is invariant under σ_t, we get the assertion.

Let M_{φ_0} denote the set of all $x \in M$ satisfying the condition in Lemma 15.8. Remark that M_{φ_0} is nothing but the algebra defined by Størmer in [18; Lemma 4.1].

LEMMA 15.9. Let ψ be a normal positive linear functional of M. If φ_0 and ψ commute with each other, that is, $|\varphi_0 + i\psi| = |\varphi_0 - i\psi|$, then ψ is invariant under the modular automorphism group σ_t.

Proof. Let $\varphi_0 + i\psi = u\omega$ be the polar decomposition of $\varphi_0 + i\psi$, that is, $\omega = |\varphi_0 + i\psi|$. By Lemma 15.7, u is a unitary operator in M. By assumption, we have

$$u^*\omega u = |\varphi_0 - i\psi| = \omega ,$$

so that we have $u\omega = \omega u$; hence $u(\varphi_0 + i\psi) = (\varphi_0 + i\psi)u$. It follows that

$$u(\varphi_0 + i\psi)u^* = \varphi_0 + i\psi$$

which means that

$$u\varphi_0 u^* = \varphi_0 \quad \text{and} \quad u\psi u^* = \psi \ .$$

Therefore, u belongs to M_{φ_0} ; hence by Lemma 15.8 u is invariant under σ_t.

Since $u\omega = \varphi_0 + i\psi$ and $u^{-1}\omega = \varphi_0 - i\psi$, we have

$$u^2(\varphi_0 - i\psi) = \varphi_0 + i\psi \ ;$$

then

$$(u^2 - 1)\varphi_0 = i(1 + u^2)\psi \ .$$

Because u^2 is in M_{φ_0}, $(u^2 - 1)\varphi_0$ is σ_t-invariant, so is $i(1 + u^2)\psi$. Therefore we have

$$(1 + u^2)(\psi - \psi\sigma_t) = 0 \ ,$$

where $\psi\sigma_t$ is defined by $\langle x, \psi\sigma_t\rangle = \langle\sigma_t(x), \psi\rangle$, $x \in M$. Let e denote the support projection $s(\psi - \psi\sigma_t)$ of $\psi - \psi\sigma_t$. Then $(1 + u^2)e = 0$ by definition of the support projection. By the equality $u^2 e = -e$, we have

$$\begin{aligned}
\varphi_0(e) - i\psi(e) &= -\langle u^2 e, \varphi - i\psi\rangle \\
&= -\langle e, (\varphi_0 - i\psi)u^2\rangle \\
&= -\langle e, u^2(\varphi_0 - i\psi)\rangle = -\langle e, \varphi_0 + i\psi\rangle \\
&= -\varphi_0(e) - i\psi(e) \ ;
\end{aligned}$$

so $\varphi_0(e) = 0$, which implies that $e = 0$. Therefore we get the invariance of ψ, that is, $\psi = \psi\sigma_t$. This completes the proof.

LEMMA 15.10. If a sequence $\{\psi_n\}$ in M_* converges to $\psi \in M_*$ in the norm topology and if a sequence $\{a_n\}$ in M converges strongly to $a \in M$, then the sequence $\{a_n\psi_n\}$ converges to $a\psi$ in the norm topology in M_*.

Proof. By the uniform boundedness Theorem, $\{a_n\}$ is bounded, so that

$$\lim_{n\to\infty} \|a_n\psi_n - a_n\psi\| \le \lim_{n\to\infty} \|a_n\|\|\psi_n - \psi\| = 0 .$$

By the inequality:

$$\|a_n\psi_n - a\psi\| \le \|a_n\psi_n - a_n\psi\| + \|a_n\psi - a\psi\| ,$$

we only have to prove $\lim \|a_n\psi - a\psi\| = 0$. Since ψ has the form $\psi = \omega_{\xi,\eta}$ for some pair ξ, η in $\mathfrak{H}$ (if we do not want to use the representation theorem of a normal linear functional ψ, then we may consider the cyclic representation of M induced by the absolute value $|\psi|$ of ψ based on the space-free property of the strong operator topology of M on its bounded part), we have the following:

$$\begin{aligned} |\langle x, a_n\psi - a\psi\rangle| &= |(x(a_n - a)\xi|\eta)| \\ &\le \|x\|\|(a_n - a)\xi\|\|\eta\| \to 0 \end{aligned}$$

as $n \to \infty$, which implies our assertion.

LEMMA 15.11. If $\psi \in M_*^+$ is invariant under σ_t, then ψ commutes with φ_0, that is,

$$|\varphi_0 + i\psi| = |\varphi_0 - i\psi| .$$

Proof. By Theorem 15.1, ψ can be uniquely represented in the form $\psi = \omega_{\xi,\xi}$ by some $\xi \in P^{\#}$. For each $\eta \in \mathfrak{A}'$, we have

$$(\Delta^{it}\xi|\eta \cdot \eta^{\flat}) = (\xi|\Delta^{-it}(\eta \cdot \eta^{\flat}))$$

$$= (\xi|(\Delta^{-it}\eta) \cdot (\Delta^{-it}\eta)^{\flat}) \geq 0 ,$$

so that $\Delta^{it}\xi$ is in $P^{\#}$. By the σ_t-invariance of ψ, we have $\psi = \omega_{\Delta^{it}\xi,\Delta^{it}\xi}$; hence we have $\Delta^{it}\xi = \xi$ by the unicity of ξ. Therefore, the h_0 defined in the proof of Lemma 15.2, the Friedrick's extension h of h_0 commutes with Δ^{it} and with every unitary operator of M'. Let

$$h = \int_0^\infty \lambda de(\lambda)$$

be the spectral decomposition of h. Put $h_n = \int_0^n \lambda de(\lambda)$. Then $\lim_{n\to\infty} h_n\eta = h\eta$ for every η in the definition domain $\mathfrak{D}(h)$. In particular, $\xi = h\xi_0 = \lim_{n\to\infty} h_n\xi_0$. Put $\psi_n = h_n\varphi_0 h_n$. Then $\lim_{n\to\infty} \|\psi - \psi_n\| = 0$. Since h_n commutes with Δ^{it}, we have

$$\psi_n = h_n\varphi_0 h_n = h_n^2\varphi_0$$

by Lemma 15.8.

For each $\eta \in \mathfrak{D}(h^2)$, we have

$$\|((1 + ih_n^2)^{-1} - (1 + ih^2)^{-1}\eta\| = \|(1 + ih_n^2)^{-1}(1 + ih^2)^{-1}(h^2 - h_n^2)\eta\|$$

$$\leq \|(h^2 - h_n^2)\eta\| \to 0$$

as $n \to \infty$. Since $\|(1 + ih_n^2)^{-1}\| \leq 1$ and $\mathfrak{D}(h^2)$ is dense in $\mathfrak{H}$, $(1 + ih_n^2)^{-1}$ converges strongly to $(1 + ih^2)^{-1}$. Therefore, by

Lemma 15.10, we have

$$\begin{aligned}(1 + ih^2)^{-1}(\varphi_0 + i\psi) &= \lim_n (1 + ih_n^2)^{-1}(\varphi_0 + i\psi_n) \\ &= \lim_n (1 + ih_n^2)^{-1}(\varphi_0 + ih_n^2\varphi_0) \\ &= \varphi_0 .\end{aligned}$$

Similarly, we have

$$(1 - ih^2)^{-1}(\varphi_0 - i\psi) = \varphi_0 .$$

Putting $v = (1 - ih^2)(1 + ih^2)^{-1}$, we get a unitary operator v in M with $v(\varphi_0 + i\psi) = \varphi_0 - i\psi$. Let

$$\varphi_0 + i\psi = u\omega, \quad \omega \in M_*^+ ,$$

be the polar decomposition of $\varphi_0 + i\psi$. Then we have

$$\varphi_0 - i\psi = vu\omega .$$

Because ω is faithful and vu is unitary as is already seen, the above equality is actually the polar decomposition of $\varphi_0 - i\psi$. Hence ω is also the absolute value of $\varphi_0 - i\psi$. This completes the proof.

Combining the above lemmas, we get the following:

THEOREM 15.2. Let ψ be a positive linear functional on M. Then the following conditions are all equivalent:

1° ψ is σ_t-invariant;

2° ψ commutes with φ_0, that is,

$$|\varphi_0 + i\psi| = |\varphi_0 - i\psi| ;$$

3° ψ has the form $\psi = \omega_{\xi,\xi}$, $\xi \in P^\#$, with $\xi = \Delta^{it}\xi$, $-\infty < t < +\infty$;

4° ψ has the form $\psi = \omega_{h\xi_0, h\xi_0}$ with h a positive self-adjoint operator affiliated with M_{φ_0}.

Now we shall determine all elements in M_*^+ majorized by a positive scalar multiple of φ_0. As seen in Theorem 15.1, if $\psi \in M_*^+$ is majorized by $\gamma\varphi_0$ for some scalar $\gamma > 0$, then ψ can be uniquely represented in the form $\psi = h\varphi_0 h$ with $h \in M_+$. [5)]

LEMMA 15.12. $\omega_{\xi,\xi}$, $\xi \in \mathfrak{H}$, is majorized by a scalar multiple $\gamma\varphi_0$ of φ_0 if and only if ξ belongs to $\mathfrak{A}'$.

The assertion is almost trivial from our preceeding study, so we omit the proof.

Therefore, our problem turns out to be the following: When does $\xi \in \mathfrak{A}$ belong to $\mathfrak{A}'$? Of course, we are required to answer the question in terms of $\pi(\xi)$ and the mdoular automorphism group σ_t rather than the trivial statement that $\xi \in \mathfrak{A} \cap \mathfrak{A}'$.

LEMMA 15.13. We have

$$(15.16) \qquad \Delta^{\frac{1}{2}}\mathfrak{A} = \mathfrak{A}' \quad \text{and} \quad \Delta^{-\frac{1}{2}}\mathfrak{A}' = \mathfrak{A} .$$

<u>Proof</u>. Recalling that $J\mathfrak{A} = \mathfrak{A}'$ and $J\mathfrak{A}' = \mathfrak{A}$ by Corollary 10.1, (15.6) follows from the equalities:

[5)] M_+ denotes the set of all positive elements in M.

$$J\xi^{\#} = \Delta^{\frac{1}{2}}\xi \quad \text{and} \quad J\xi^{b} = \Delta^{-\frac{1}{2}}\xi \ .$$

Therefore, we have at once the following:

LEMMA 15.14. $\xi \in \mathfrak{A}$ belongs to $\mathfrak{A}'$ if and only if $\Delta^{\frac{1}{2}}\xi^{\#}$ belongs to $\mathfrak{A}$ again.

Therefore, by the equality:

$$\pi(\Delta^{\frac{1}{2}}\xi) = \Delta^{\frac{1}{2}}\pi(\xi)\Delta^{-\frac{1}{2}} \ ,$$

we have the following result:

LEMMA 15.15. $a\varphi_0 a^*$, $a \in M$, is majorized by $\gamma\varphi_0$ for some $\gamma > 0$, if and only if $\Delta^{\frac{1}{2}}a^*\Delta^{-\frac{1}{2}}$ is bounded. In particular, $a\varphi_0 a^* \leq \varphi_0$ if and only if $\|\Delta^{\frac{1}{2}}a^*\Delta^{-\frac{1}{2}}\| \leq 1$, equivalently, $\|\Delta^{-\frac{1}{2}}a\Delta^{\frac{1}{2}}\| \leq 1$.

Put $a = \pi(\xi)$, $\xi \in \mathfrak{A}$. Let $\mathfrak{A}_0$ be the modular Hilbert algebra constructed in §10 from $\mathfrak{A}$. For each pair η, ζ in $\mathfrak{A}_0$, consider a function $F_{\eta,\zeta}(z)$ of a complex variable defined by:

$$F_{\eta,\zeta}(z) = ((\Delta^{z}\xi)\eta|\zeta) \ .$$

Then by the equality:

$$(\Delta^{z}\xi\eta|\zeta) = (\xi\Delta^{-z}(\eta\zeta^{b})|\xi_0) \ ,$$

$F_{\eta,\zeta}(z)$ can be extended to an analytic function on the complex plane C. For each real number t, we have

$$|F_{\eta,\zeta}(it)| = |((\Delta^{it}\xi)\eta|\zeta)|$$

$$\leq \|\pi(\Delta^{it}\xi)\|\|\eta\|\|\zeta\| = \|a\|\|\eta\|\|\zeta\| \ ;$$

$$|F_{\eta,\zeta}(-\tfrac{1}{2} + it)| = |((\Delta^{-\frac{1}{2}+it}\xi)\eta|\zeta)|$$

$$\leq \|\pi(\Delta^{it}\Delta^{-\frac{1}{2}}\xi)\|\|\eta\|\|\zeta\|$$

$$= \|\pi(\Delta^{-\frac{1}{2}}\xi)\|\|\eta\|\|\zeta\| \leq \|\Delta^{-\frac{1}{2}}a\Delta^{\frac{1}{2}}\|\|\eta\|\|\zeta\| \ .$$

Therefore, by the theorem of Phragmen-Lindelöf, we have

$$|F_{\eta,\zeta}(s + it)| \leq \|a\|^{1-2s}\|\Delta^{-\frac{1}{2}}a\Delta^{\frac{1}{2}}\|^{2s}\|\eta\|\|\zeta\|$$

for $-\frac{1}{2} \leq s \leq 0$. Now take an arbitrary pair η, ζ in $\mathfrak{H}$, then there exist two sequences $\{\eta_n\}$ and $\{\zeta_n\}$ in $\mathfrak{A}_0$ with $\eta = \lim \eta_n$ and $\zeta = \lim \zeta_n$. Then the sequence $\{F_{\eta_n,\zeta_n}(z)\}$ of analytic functions converges uniformly on the strip $-\frac{1}{2} \leq \text{Re } z \leq 0$ to the function $F_{\eta,\zeta}(z)$. Therefore, $F_{\eta,\zeta}(z)$ is continuous on and analytic in the strip $0 \leq \text{Im } z \leq \frac{1}{2}$. On the boundary, we have

$$F_{\eta,\zeta}(it) = ((\Delta^{it}\xi)\eta|\zeta) \ ;$$

$$F_{\eta,\zeta}(-\tfrac{1}{2} + it) = ((\Delta^{it}\Delta^{-\frac{1}{2}}\xi)\eta|\zeta) \ .$$

Thus, the function: $t \to \sigma_t(a) = \pi(\Delta^{it}\xi)$ is extended analytically in and continuously on the strip $-\frac{1}{2} \leq \text{Im } z \leq 0$, where the strong operator topology is considered in M. After all, we get the following criteria for $\psi = a\varphi_0 a^*$ to be majorized by some scalar multiple of φ_0:

THEOREM 15.3. For each $\xi \in \mathfrak{H}$, the following statements are equivalent:

A1° $\omega_{\xi,\xi}$ is majorized by φ_0, i.e. $\omega_{\xi,\xi} \leq \varphi_0$;

A2° ξ belongs to $\mathfrak{A}'$ and $\|\pi'(\xi)\| \leq 1$;

A3° $\Delta^{\frac{1}{2}}\xi^{\#}$ belongs to $\mathfrak{A}$ and $\|\pi(\Delta^{\frac{1}{2}}\xi^{\#})\| \leq 1$, whenever $\xi \in \mathfrak{A}$;

A4° $0 \leq \pi(\xi) \leq 1$ and ξ belongs to $\mathfrak{A}'$ with $\|\pi'(\xi)\| \leq 1$ whenever $\xi \in \mathfrak{A} \cap P^{\#}$.

For each $a \in M$, the following statements are equivalent:

B1° $\psi = a\varphi_0 a^*$ is majorized by φ_0;

B2° $\Delta^{-\frac{1}{2}}a\Delta^{\frac{1}{2}}$ is bounded and $\|\Delta^{-\frac{1}{2}}a\Delta^{\frac{1}{2}}\| \leq 1$;

B3° The function: $t \to \sigma_t(a)$ is extended to a function $\sigma_z(a)$ which is analytic in and continuous on the strip: $0 \leq \operatorname{Im} z \leq \frac{1}{2}$ and $\|\sigma_{t+\frac{1}{2}i}(a)\| \leq 1$.

In §13, we have seen the unicity of the modular automorphism group associated with a faithful normal positive linear functional. Applying the above "Radon-Nikodym" arguments, we shall show the converse unicity. Namely, if we fix a one parameter automorphism group σ_t of a von Neumann algebra M, then the normal positive linear functional of M satisfying the KMS-boundary condition with respect to σ_t is unique up to the center of M.

THEOREM 15.4. Let φ_0 be a faithful normal positive linear function of a von Neumann algebra M with the modular automorphism group σ_t. If a normal positive linear functional ψ of M satisfies the KMS-boundary condition with respect to σ_t for $\beta = 1$, then ψ has the form $\psi = \omega_{h\xi_0, h\xi_0}$ with h a positive self-adjoint operator affiliated with the center Z of M.

In particular, if M is a factor, then ψ is a scalar multiple of φ_0.

<u>Proof</u>. By Theorem 13.3, the support projection e of ψ is central, so that we may assume that ψ is faithful, considering ψ on Me if necessary. Since ψ is σ_t-invariant, Theorem 15.2 assures that ψ has the form $\psi = \omega_{h\xi_0, h\xi_0}$ with h a positive self-adjoint operator affiliated with M_{φ_0}. Remark that the range projection of h is the support projection of ψ. Hence h has the dense range. Let

$$h = \int_0^\infty \lambda de(\lambda)$$

be the spectral decomposition of h. Then all projections $\{e(\lambda)\}$ are in M_{φ_0}. Put

$$h_n = e\left(\frac{1}{n}\right) + \int_{1/n}^{n} \lambda de(\lambda) + (1 - e(n)) ;$$

$$\psi_n(x) = \varphi_0(h_n x h_n) = \varphi_0(x h_n^2), \quad n = 1,2,\dots .$$

Since $h_n\xi_0$ converges strongly $h\xi_0$ as $n \to \infty$, ψ_n converges to ψ with respect to the norm topology in M_*. Define a strongly continuous one parameter unitary group $u_n(t)$ in M with bounded infinitesimal generator $2 \log h_n$ (so that analytic) by:

$$u_n(t) = h_n^{2it} = \exp(2it \log h_n), \quad n = 1,2,\dots .$$

Then for each fixed t, $u_n(t)$ converges strongly to the unitary operator $u(t) = h^{2it}$. For each pair x, y in M and $n = 1,2,\ldots$, define a function $f_n(t,z)$ of two variables (t,z) by:

$$f_n(t,z) = \psi_n(h_n^{2iz}\sigma_t(x)h_n^{-2iz}y) .$$

Then f_n can be extended to a function $F_n(w,z)$ analytic in and continuous on the region $\{(w,z);\ 0 \leq \operatorname{Im} w \leq 1,\ z \in C\}$. Now we shall compute the boundary value $F_n(t + i,\ z + i)$ as follows: From the equality:

$$\psi_n(h_n^{2iz}\sigma_t(x)h_n^{-2iz}y) = \varphi_0(\sigma_t(x)h_n^{-2iz}yh_n^{2iz}h_n^2) ,$$

it follows that

$$\begin{aligned} F_n(t + i,z) &= \varphi_0(h_n^{-2iz}yh_n^{2i(z-i)}\sigma_t(x)) \\ &= \varphi_0(yh_n^{2i(z-i)}\sigma_t(x)h_n^{-2iz}) \\ &= \varphi_0(yh_n^{2i(z-i)}\sigma_t(x)h_n^{-2i(z-i)}h_n^2) \\ &= \psi_n(yh_n^{2i(z-i)}\sigma_t(x)h_n^{-2i(z-i)}) . \end{aligned}$$

Hence we have

$$F_n(t + i,\ z + i) = \psi_n(yh_n^{2iz}\sigma_t(x)h_n^{-2iz}) .$$

Therefore the one parameter automorphism group σ_t^n defined by:

$$\sigma_t^n(x) = u_n(t)\sigma_t(x)u_n(t)^{-1}$$

is the modular automorphism group associated with ψ_n. Define a one parameter automorphism group σ_t' of M by

$$\sigma_t'(x) = u(t)\sigma_t(x)u(t)^{-1} = h^{2it}\sigma_t(x)h^{-2it} .$$

For each pair x, y in M, define functions f(t) and g(t) by:

$$f(t) = \psi(\sigma_t'(x)y) \text{ and } g(t) = \psi(y\sigma_t'(x)) .$$

Then functions defined by:

$$f_n(t) = \psi_n(\sigma_t^n(x)y) \text{ and } g_n(t) = \psi_n(y\sigma_t^n(x))$$

converge f(t) and g(t) respectively. The function $F_n(z)$ defined by $F_n(z,z)$, where $F_n(w,z)$ is the one defined above, is analytic in and continuous on the strip: $0 \leq \text{Im } z \leq 1$ and has boundary values: $f_n(t) = F_n(t)$ and $g_n(t) = F_n(t + i)$. Therefore, the sequence $\{F_n(z)\}$ is uniformly bounded and converges on the boundary, so that $\{F_n(z)\}$ converges uniformly to a function $F(z)$, on every compact subset of the open strip: $0 < \text{Im } z < 1$. Hence $F(z)$ is analytic in and continuous on the strip: $0 \leq \text{Im } z \leq 1$ and has boundary value: $F(t) = f(t)$ and $F(t + i) = g(t)$. Therefore, σ_t' is the modular automorphism group associated with ψ, so that we have $\sigma_t' = \sigma_t$ by assumption, which means that $u(t)$ is in the center. Therefore h must be affiliated with the center.

Remark. Our Theorem 15.4 tells us that the set of all states of a C^*-algebra satisfying the KMS-boundary condition with respect to a fixed one-parameter automorphism group and a fixed β forms a (Choquet) simplex in the sense of [25].

REFERENCES

1. H. Araki: Multiple time analyticity of a quantum statistical state satisfying the KMS boundary condition, to appear.
2. H. Araki and H. Miyata: On KMS boundary condition, to appear.
3. J. Dixmier: Les algèbres d'opérateurs dans l'espace hilbertien, Gauthier-Villars, Paris, 2$^{\acute{e}}$ edition, (1969).
4. J. Dixmier: Algèbres quasi-unitaires, Comm. Math. Helv., 26 (1952), 275-322.
5. J. Dixmier: Formes lineaires sur un anneau d'opérateurs, Bull. Soc. Math. Fr., 81 (1953), 9-39.
6. N. Dunford and J. T. Schwartz: Linear operators II, Interscience Publications, New York, (1963).
7. H. Dye: The Radon-Nikodym theorem for finite rings of operators, Trans. Amer. Math. Soc., 72 (1952), 243-280.
8. R. Haag, N. Hugenholtz and M. Winnink: On the equilibrium states in quantum statistical mechanics, Comm. Math. Phys., 5 (1967), 215-236.
9. E. Hille and R. S. Phillips: Functional analysis and semi-groups, Amer. Math. Colloq. Pub., 31 (1957).
10. N. M. Hugenholtz: On the factor type of equilibrium states in quantum statistical mechanics, Comm. Math. Phys., 6 (1967), 189-193.
11. N. M. Hugenholtz and J. D. Wieringa: On locally normal states in quantum statistical mechanics, to appear.
12. D. Kastler, J. C. T. Pool and E. Thue Poulsen: Quasi-unitary algebras attached to temperature states in statistical mechanics - a comment on the work of Haag, Hugenholtz and Winnink, to appear.

13. F. J. Murray and J. von Neumann: On rings of operators, Ann. Math., 37 (1936), 116-229.

14. J. von Neumann: On rings of operators III, Ann. Math., 41 (1940), 94-161.

15. S. Sakai: The theory of W^*-algebras, Lecture Notes, Yale Univ., (1962).

16. S. Sakai: A Radon-Nikodym theorem in W^*-algebras, Bull. Amer. Math. Soc., 71 (1965), 149-151.

17. I. E. Segal: A non-commutative extension of abstract integration, Ann. Math., 57 (1953), 401-457.

18. E. Størmer: States and invariant maps of operator algebras, to appear in J. Functional Analysis.

19. M. Takesaki: On the conjugate space of operator algebra, Tôhoku Math. J., 10 (1958), 194-203.

20. M. Takesaki: A characterization of group algebras as a converse of Tannaka-Stinespring-Tatsuuma duality theorem, to appear in Amer. J. Math.

21. M. Tomita: Quasi-standard von Neumann algebras, mimeographed note, (1967).

22. M. Tomita: Standard forms of von Neumann algebras, the Vth functional analysis symposium of the Math. Soc. of Japan, Sendai, (1967).

23. B. J. Vowden: A new proof in the spatial theory of von Neumann algebras, J. London Math. Soc., 44 (1969), 429-432.

24. R. V. Kadison: Transformations of states in operator theory and dynamics, Topology, 3 (1965), 177-198.

25. R. Phelps: Lectures on Choquet's theorem, von Nostrand, Princeton, (1966).

26. L. Pukanszky: On the theory of quasi-unitary algebras, Acta Sci. Math., 17 (1955), 103-121.

This paper was supported in part by the National Science Foundation Grant #GP-7683.

UCLA

AND

TÔHOKU University, Japan

Offsetdruck: Julius Beltz, Weinheim/Bergstr.